WiMAX

Standards and Security

The
WiMAX
Handbook

WiMAX: Technologies, Performance Analysis, and QoS
ISBN 9781420045253

WiMAX: Standards and Security
ISBN 9781420045237

WiMAX: Applications
ISBN 9781420045474

The WiMAX Handbook
Three-Volume Set
ISBN 9781420045350

WiMAX

Standards and Security

Edited by
SYED AHSON
MOHAMMAD ILYAS

CRC Press
Taylor & Francis Group
Boca Raton   London   New York

CRC Press is an imprint of the
Taylor & Francis Group, an informa business

CRC Press
Taylor & Francis Group
6000 Broken Sound Parkway NW, Suite 300
Boca Raton, FL 33487-2742

International Standard Book Number-10: 1-4200-4523-7 (Hardcover)
International Standard Book Number-13: 978-1-4200-4523-9 (Hardcover)

Library of Congress Cataloging-in-Publication Data

Ahson, Syed.
   WiMAX : standards and security / Syed Ahson and Mohammad Ilyas.
      p. cm.
   Includes bibliographical references and index.
   ISBN 978-1-4200-4523-9 (alk. paper)
    1. Wireless communication systems. 2. Broadband communication systems.
  3. IEEE 802.16 (Standard) I. Ilyas, Mohammad, 1953- II. Title.

TK5103.2.A432 2008
621.384--dc22   2007012500

Visit the Taylor & Francis Web site at
http://www.taylorandfrancis.com

and the CRC Press Web site at
http://www.crcpress.com

Contents

Part II   Security

Preface

The demand for broadband services is growing exponentially. Traditional solutions that provide high-speed broadband access use wired access technologies, such as traditional cable, digital subscriber line, Ethernet, and fiber optic. It is extremely difficult and expensive for carriers to build and maintain wired networks, especially in rural and remote areas. Carriers are unwilling to install the necessary equipment in these areas because of little profit and potential. WiMAX will revolutionize broadband communications in the developed world and bridge the digital divide in developing countries. Affordable wireless broadband access for all is very important for a knowledge-based economy and society. WiMAX will provide affordable wireless broadband access for all, improving quality of life thereby leading to economic empowerment.

Broadband wireless access technical solutions and products have been available for some time. These technologies have primarily focused on providing high data rate connectivity wirelessly between fixed stationary sites. These technical solutions are proprietary in nature and suffer from poor interoperability with other broadband wireless access products and have a high cost due to the lack of economy of scale. High-speed wireless services have already achieved great success in local area networks with the IEEE 802.11 standard and Wi-Fi certified products.

The IEEE 802.16 BWA technology family, referred to as Worldwide Interoperability for Microwave Access intends to provide a standardized BWA solution. The IEEE Standards Board established the IEEE 802.16 working group in 1999 to prepare formal specifications for global deployment of broadband Wireless Metropolitan Area Networks, officially called WirelessMAN. The WiMAX Forum, created in 2003, is promoting the commercialization of IEEE 802.16 and the European Telecommunications Standard Institute's high-performance radio MAN. The IEEE 802.16 specifications continue to evolve and expand in capabilities in support of the evolving vision of WiMAX usage and deployment. The IEEE 802.16e system called Mobile WiMAX has been standardized to add user mobility to the original IEEE 802.16 system.

WiMAX has a strong base of standardization and industry support that provides a strong evolutionary path of its capabilities. WiMAX competes with IEEE 802.11-based WLAN technology, broadband residential Internet technologies such as digital subscriber line and cable and third-generation cellular technologies. WiMAX is the next step in the mobile technology evolution path. WiMAX will broaden wireless access to metropolitan area networks. WiMAX offers numerous advantages, such as improved performance and robustness, end-to-end IP-based networks, secure mobility and broadband speeds for voice, data, and video, support for fixed and mobile systems, efficient and adaptive coding and modulation techniques, scalable channel sizes, subchannelization schemes, multiple-input-multiple-output antenna systems, and quality of service. WiMAX enables wireless broadband access anywhere, anytime, and on virtually any device.

The WiMAX handbook provides technical information about all aspects of WiMAX. The areas covered in the handbook range from basic concepts to research-grade material including future directions. The WiMAX handbook captures the current state of wireless local area networks, and serves as a source of comprehensive reference material on this subject. The WiMAX handbook consists of three volumes: *WiMAX: Applications; WiMAX: Standards and Security*; and *WiMAX: Technologies, Performance Analysis, and QoS*. It has a total of 32 chapters authored by experts from around the world. *WiMAX: Standards and Security* includes 12 chapters authored by 22 experts.

Chapter 1 (The Emerging Wireless Internet Architecture: Competing and Complementary Standards to WiMAX Technology) describes other wireless networking technologies that complement and compete with WiMAX technologies. This chapter provides an overview of the most prevalent current technologies in use today, as well as a description of the similarities and differences compared to WiMAX.

Chapter 2 (IEEE 802.16 Standards and Amendments) examines the pros and cons of standardized versus proprietary solutions for wireless broadband access. An overview of WiMAX standards and amendments (IEEE 802.16-2001, IEEE 802.16b, IEEE 802.16c, IEEE 802.16d, IEEE 802.16-2004, IEEE 802.16e-2005, IEEE 802.16f, IEEE 802.16g, IEEE 802.16h, IEEE 802.16fi, and IEEE 802.16j) is presented. Key WiMAX technologies such as physical layer, medium access control layer, convergence sublayer, common part sublayer, point-to-multipoint and mesh mode, privacy sublayer, quality of service support, handover support, and power management are described in detail.

Chapter 3 (MAC Layer Protocol in WiMAX Systems) reviews the functions and features of the core medium access control protocol of the WiMAX systems including the point-to-multipoint topology and mesh topology. The fundamental part of the medium access control protocol of the WiMAX systems is summarized and presented.

Chapter 4 (Scheduling and Performance Analysis of QoS for IEEE 802.16 Broadband Wireless Access Network) presents an architecture and its implementation of admission control and job scheduling based on the quality-of-service requirements of IEEE 802.16. This chapter presents the concept and requirements of quality of service as specified in the IEEE 802.16 standard, along with an architecture to implement quality of service in a simulation model.

Chapter 5 (Propagation and Performance) presents carriers' perspectives for wireless services like fixed WiMAX access. This chapter presents various aspects of propagation and performance for WiMAX radio systems; it reviews WiMAX radio system parameters such as link budgets, presents relevant propagation models, and finally, analyzes system throughput and performance for a typical suburban area.

Chapter 6 (Mobility Support for IEEE 802.16e System) discusses the main mobility functions defined in the IEEE 802.16e standard: power-saving mechanism, handover operation, paging, and location update. Power-saving classes of type I, type II, and type III are discussed in great detail. Network topology acquisition, basic handover operation, macro-diversity handover, and fast base station switching are examined. Basic paging operation, location update, and network reentry from idle mode are described.

Chapter 7 (Measured Signal-Aware Mechanism for Fast Handover in WiMAX Networks) describes how to use a measured signal-aware mechanism to aid speeding up WiMAX handover procedures. A measured signal-aware mechanism for a base station initialized predicted handover scheme is investigated, which centralized a monitor-moving mobile subscriber station and prepared a CDMA ranging code of boundary mobile subscriber stations beforehand.

Chapter 8 (802.16 Mesh Networking) presents an overview of the 802.16 mesh protocol with a specific focus on the networking aspects of the protocol. Addressing assignments for IEEE 802.16 mesh networks that allow the network layer to take advantage of quality of service provided by IEEE 802.16 mesh protocol is proposed. An overview of the security infrastructure of IEEE 802.16 mesh networks and their flaws is presented. An end-to-end security scheme that simplifies the design of IEEE 802.16 mesh routers is proposed.

Chapter 9 (WiMAX Testing) surveys the testing and certification processes used for WiMAX products. This chapter describes the general framework used for conformance and interoperability testing for the WiMAX technology. An overview of generic test equipment, test environments, and scenarios used for WiMAX certification testing is described. It also describes the WiMAX certification process and testing scenarios at the recently held WiMAX Forum "Plugfest" events.

Chapter 10 (An Overview of WiMAX Security) presents an overview of the security aspects of IEEE 802.16. Unified modeling language class and sequence diagrams are used to describe architectural aspects. These are conceptual diagrams, intended to define the information in each unit and do not reflect implementation details. This chapter presents a high-level overview that can be read before getting into the details of the standard.

Chapter 11 (Privacy and Security in WiMAX Networks) presents an overview of WiMAX security features. Primary, static, and dynamic security associations, contents of data security association, and contents of authorization security association are described in detail. Hashed message authentication codes, X.509 certificates, and the extensible authentication protocol are reviewed. Aspects of privacy and key management protocol such as authorization and authorization key exchange, and traffic encryption key exchange are examined.

Chapter 12 (WiMAX Security: Privacy Key Management) presents a comprehensive overview of security issues encountered in WiMAX, including security challenges, user authentication, key exchanges, as well as data encryption through the fixed and mobile WiMAX channels. This chapter focuses on the privacy and key management protocols that play an important role in securing connection and transmission across broadband wireless access.

The targeted audience for the handbook includes professionals who are designers and planners for WiMAX networks, researchers (faculty members and graduate students), and those who would like to learn about this field.

The handbook has the following specific salient features:

- To serve as a single comprehensive source of information and as reference material on WiMAX networks.
- To deal with an important and timely topic of emerging communication technology of today, tomorrow, and beyond.
- To present accurate, up-to-date information on a broad range of topics related to WiMAX networks.
- To present material authored by the experts in the field.
- To present the information in an organized and well-structured manner.

Although the handbook is not precisely a textbook, it can certainly be used as a textbook for graduate and research-oriented courses that deal with WiMAX. Any comments from the readers will be highly appreciated.

Many people have contributed to this handbook in their unique ways. The first and foremost group that deserves immense gratitude is the group of highly talented and skilled researchers who have contributed 32 chapters to this handbook. All of them have been extremely cooperative and professional. It has also been a pleasure to work with Nora Konopka, Helena Redshaw, Jessica Vakili, and Joette Lynch of Taylor & Francis and we are extremely gratified for their support and professionalism. Our families have extended their unconditional love and strong support throughout this project and they all deserve very special thanks.

Syed Ahson
Plantation, FL, USA

Mohammad Ilyas
Boca Raton, FL, USA

Editors

Syed Ahson is a senior staff software engineer with Motorola Inc. He has extensive experience with wireless data protocols (TCP/IP, UDP, HTTP, VoIP, SIP, H.323), wireless data applications (Internet browsing, multimedia messaging, wireless e-mail, firmware over-the-air update), and cellular telephony protocols (GSM, CDMA, 3G, UMTS, HSDPA). He has contributed significantly in leading roles toward the creation of several advanced and exciting cellular phones at Motorola. Prior to joining Motorola, he was a senior software design engineer with NetSpeak Corporation (now part of Net2Phone), a pioneer in VoIP telephony software.

Syed is a coeditor of the *Handbook of Wireless Local Area Networks: Applications, Technology, Security, and Standards* (CRC Press, 2005). Syed has authored "Smartphones" (International Engineering Consortium, April 2006), a research report that reflects on smartphone markets and technologies. He has published several research articles in peer-reviewed journals and teaches computer engineering courses as adjunct faculty at Florida Atlantic University, Florida, where he introduced a course on smartphone technology and applications. Syed received his BSc in electrical engineering from Aligarh University, India in 1995 and an MS in computer engineering in July 1998 at Florida Atlantic University, Florida.

Dr. Mohammad Ilyas received his BSc in electrical engineering from the University of Engineering and Technology, Lahore, Pakistan, in 1976. From March 1977 to September 1978, he worked for the Water and Power Development Authority, Pakistan. In 1978, he was awarded a scholarship for his graduate studies and he completed his MS in electrical and electronic engineering in June 1980 at Shiraz University, Shiraz, Iran. In September 1980, he joined the doctoral program at Queen's University in Kingston, Ontario, Canada. He completed his PhD in 1983. His doctoral research was about switching and flow control techniques in computer communication networks. Since September 1983, he has been with the College of Engineering and Computer Science at Florida Atlantic University, Boca Raton, Florida, where he is currently associate dean for research and industry relations. From 1994 to 2000, he was chair of the Department of Computer Science and Engineering. From July 2004 to September 2005, he served as interim associate vice president for research and graduate studies. During the 1993–1994 academic year, he was on his sabbatical leave with the Department of Computer Engineering, King Saud University, Riyadh, Saudi Arabia.

Dr. Ilyas has conducted successful research in various areas including traffic management and congestion control in broadband/high-speed communication networks, traffic characterization, wireless communication networks, performance modeling, and simulation. He has published one book, eight handbooks, and over 150 research articles. He has supervised 11 PhD dissertations and more than 37 MS theses to completion. He has been a consultant to several national and international organizations. Dr. Ilyas is an active participant in several IEEE technical committees and activities.

Dr. Ilyas is a senior member of IEEE and a member of ASEE.

Contributors

Rana Ejaz Ahmed
Department of Computer
   Engineering
American University of Sharjah
Sharjah, United Arab Emirates

Najah Abu Ali
College of Information Technology
United Arab Emirates University
Al-Ain, United Arab Emirates

Nirwan Ansari
Department of Electrical and
   Computer Engineering
New Jersey Institute of Technology
Newark, New Jersey

Jack L. Burbank
The Johns Hopkins University
   Applied Physics Laboratory
Laurel, Maryland

Jenhui Chen
Department of Computer Science
   and Information Engineering
Chang Gung University
Taiwan, Republic of China

Dong-Ho Cho
Korea Advanced Institute of Science
   and Technology
Daejeon, Republic of Korea

Hyun-Ho Choi
Korea Advanced Institute of Science
   and Technology
Daejeon, Republic of Korea

Petar Djukic
Department of Electrical and
   Computer Engineering
University of Toronto
Toronto, Ontario, Canada

Eduardo B. Fernandez
Department of Computer Science
   and Engineering
Florida Atlantic University
Boca Raton, Florida

Nolan Glore
Bradley Department of Electrical
   and Computer Engineering
Virginia Polytechnic Institute and
   State University
Blacksburg, Virginia

Hossam S. Hassanein
School of Computing
Queen's University
Kingston, Ontario, Canada

Edwin Hou
Department of Electrical and
   Computer Engineering
New Jersey Institute of Technology
Newark, New Jersey

William T. Kasch
The Johns Hopkins University
   Applied Physics Laboratory
Laurel, Maryland

Yuanqiu Luo
NEC Laboratories America, Inc.
Princeton, New Jersey

Maode Ma
School of Electrical and Electronic
    Engineering
Nanyang Technological University
Singapore

Amitabh Mishra
Bradley Department of Electrical
    and Computer Engineering
Virginia Polytechnic Institute and
    State University
Blacksburg, Virginia

Thomas Schwengler
Qwest Communications
Denver, Colorado

Shahrokh Valaee
Department of Electrical and
    Computer Engineering
University of Toronto
Toronto, Ontario, Canada

Michael VanHilst
Department of Computer Science
    and Engineering
Florida Atlantic University
Boca Raton, Florida

Chih-Chieh Wang
Department of Electrical
    Engineering
Chang Gung University
Taiwan, Republic of China

James T. Yu
DePaul University
Chicago, Illinois

Chao Zhang
Department of Electrical and
    Computer Engineering
New Jersey Institute of Technology
Newark, New Jersey

Yan Zhang
Simula Research Laboratory
Lysaker, Norway

Part I
Standards

1

The Emerging Wireless Internet Architecture: Competing and Complementary Standards to WiMAX Technology

William T. Kasch and Jack L. Burbank

CONTENTS

1.1 Introduction

Until the year 2000, users of the Internet accessed its contents primarily through wired, fixed infrastructure sites (e.g., universities, home dial-up connections, and corporate and government facilities). However, technology has evolved such that a significant number of users today access Internet services wirelessly. This "access revolution" has gone hand-in-hand with the increasing usage of laptop computers and smaller mobile wireless devices such as cellular telephones and RIM BlackBerry™ devices. The cumulative result has created an information-centric society where users rely on network services in most aspects of their day-to-day life. The emerging wireless Internet architecture aims to continue the access revolution by supporting an increasing number of users at increased data rates, such that the user experience is similar to the experience from a wired, high-speed connection. A variety of wireless technologies have been proposed, both in standards organizations and by industry consortiums, to enable wireless network access. This chapter discusses some of the most popular technologies available today, those that are expected to be available in the future, and how these technologies may compete or compliment WiMAX technology.

1.2   The IEEE 802 Standards Family

Figure 1.1 shows the structure of the IEEE 802 standards family of ratified technologies. IEEE 802 primarily focuses on the physical (PHY) and media access (MAC) layer specifications of the 7-layer open systems interconnection (OSI) model context. Such standards in the IEEE 802 family include the IEEE 802.3 (wired Ethernet) standard, IEEE 802.1 (management) standard, IEEE 802.5 (token ring) standard, and the widely deployed IEEE 802.11 (wireless local area networks or WLAN) standard. WiMAX technology is primarily based on the IEEE 802.16 (wireless metropolitan area networks or WMAN) standard, while Bluetooth and ZigBee share similarities to some elements within the IEEE 802.15 standard.

With the recent success and wide adoption of IEEE 802.11 WLAN technology, IEEE 802 has developed other standards that aim to take the emerging wireless Internet architecture even further. IEEE 802.16 technology is aimed at providing high-speed metropolitan area level access (similar to cellular infrastructure but advertised as a fraction of the cost). The IEEE 802.16e standard aims to provide WMAN access to mobile users moving at vehicular speeds.

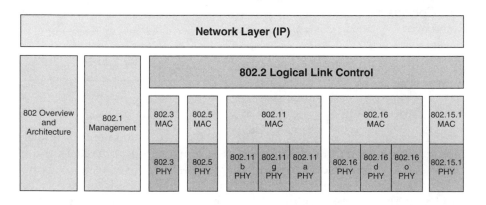

FIGURE 1.1
IEEE 802 standards family.

Each particular technology that has been released from the IEEE 802 group is focused on a narrow set of usage cases (e.g., range, mobility speed, and mesh networking) but deployments in the marketplace have often pushed technologies further (e.g., range extension of IEEE 802.11).

A notional view of an IEEE 802 wireless Internet architecture is presented in Figure 1.2. Here, an IEEE 802.16 network is deployed to enable connectivity across a large area (on the order of a city, say around 100 km^2). Within the IEEE 802.16 network, users (known as subscriber stations or SS) may access base stations (BS) directly or gateways that bridge connections to other technologies (e.g., cellular, and wired infrastructure) may be employed. In the figure, three locations are shown where connections are bridged between the IEEE 802.16 network and IEEE 802.11 access point networks. Here, the IEEE 802.16 network acts as a backhaul network while the IEEE 802.11 networks provide localized coverage to individual users or other gateway nodes (on the order of a city block, perhaps 10 km^2). The gateway nodes shown in the IEEE 802.11 network bridge connections to IEEE 802.15 wireless personal area networks (WPANs). These IEEE 802.15 networks may provide micro-local coverage (on the order of 10 ft^2) to devices such as cellular telephones, computer mice, or household appliances.

While the WiMAX Forum has been formed to promote IEEE 802.16, certified products are just now being released into the marketplace. WiMAX

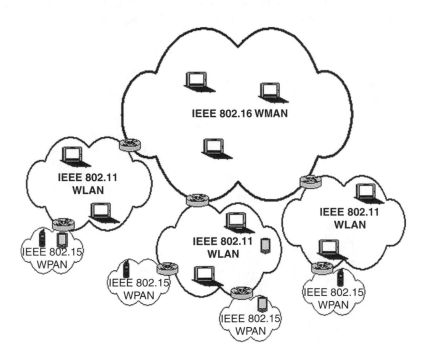

FIGURE 1.2
Notional IEEE 802 wireless internet architecture.

technology proliferation is expected to increase substantially as network service providers adopt the technology. Sprint Corporation announced in 2006 that it plans to deploy a full WiMAX network across its entire U.S. coverage area to be operational in 2007. Other corporations across the world have also announced plans to increasingly deploy WiMAX technology, especially to underserved areas such as developing countries with limited infrastructure options. Furthermore, Intel's announcement to support WiMAX as part of its wireless networking chipset in future laptop computers has further solidified WiMAX as a likely technology candidate for the next generation of wireless network-enabled devices.

1.2.1   IEEE 802.11

Of the wireless networking technologies specified by IEEE 802, IEEE 802.11 (Figure 1.3) has experienced the widest deployment to date with hundreds of thousands of IEEE 802.11 networks deployed all over the world. IEEE 802.11 supports data rates from 1 up to 54 Mbps using a variety of modulation and coding methods. IEEE 802.11b operates using a direct-sequence spread spectrum (DSSS) waveform supporting data rates up to 11 Mbps, while IEEE 802.11g uses an orthogonal frequency division multiplexing (OFDM) waveform supporting data rates up to 54 Mbps. Both IEEE 802.11b and IEEE 802.11g operate in the 2.4 GHz industrial, scientific, and medical (ISM) band, while IEEE 802.11a operates in the 5 GHz Unlicensed National Information

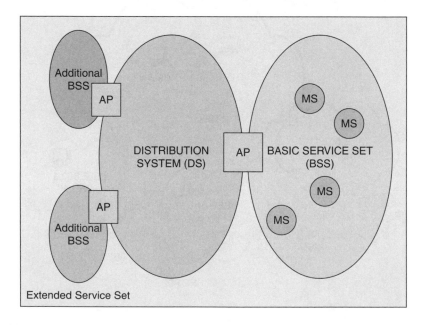

FIGURE 1.3
IEEE 802.11 network architecture.

Infrastructure (UNII) band. IEEE 802.11a uses the OFDM waveform specified in IEEE 802.11g for data rates up to 54 Mbps, albeit at lower transmit powers (around a maximum of 20 mW for IEEE 802.11a compared to a maximum of 100 mW for IEEE 802.11g). More information on these standards can be found in Refs. 1–4.

The basic service set (BSS) is the foundation of an 802.11 network. The BSS is a group of stations that communicate with one another. These communications take place in the basic service area (BSA). A station within the BSA can communicate with other members of the BSS. There are two types of BSS': *ad hoc* (or independent) and infrastructural. An *ad hoc* BSS, also known as an independent basic service set (IBSS), is one in which stations communicate directly with one another. IBSS' are typically short-lived in nature and are, thus, referred to as *ad hoc*. These are the least common types of 802.11 networks within the commercial domain. An infrastructural BSS is one in which all communications take place through the access point (AP) within that BSS. This is the most common type of 802.11 network within the commercial domain. Multiple BSS' can be interconnected into an extended service set (ESS). An ESS is formed by chaining BSS' together with a backbone network. The 802.11 does not specify the backbone network, but rather that this backbone network provides a certain set of services. From the perspective of the logical link control (LLC) sublayer that resides between the 802.11 MAC layer and the IP network layer, an ESS appears identical to a larger BSS (i.e., the concept of BSS versus ESS is transparent to the higher LLC sublayer). Figure 1.3 depicts the 802.11 network architecture from an infrastructural mode perspective.

An ESS or BSS is identified by its service set identity (SSID). The SSID is a 0- to 32-byte identifier that is typically assigned a human-readable American Standard Code for Information Interchange (ASCII) character string. As a result, it is alternatively known as the 802.11 network name. The first thing a mobile station (MS) wishing to join an 802.11 network must do is detect the presence of the network. There are two methods by which this can be accomplished: passive and active. In the passive case, the MS scans all frequency channels listening for the presence of network beacons, which are periodically transmitted by the stations of the network to announce their presence. These beacons contain essential information about that network, such as its SSID. The station can then begin the authentication and association procedures required to join the network. In the active case, the MS begins transmitting probes with the SSID of the network it wishes to join and then waits for a response from the probes. Upon receipt of a probe response, the MS can then begin joining the network. In fact, this active method is required if SSID broadcast is suppressed for security purposes.

1.2.2   IEEE 802.20

The IEEE 802.20 standard defines a wireless broadband networking technology operating in bands below 3.5 GHz with data rates around 1 Mbps.

IEEE 802.20 aims to operate in ranges up to 15 km, supporting vehicular motion up to 250 km/h (train speeds). Activities of this group were suspended on June 8, 2006, but a path forward was established on September 15, 2006 by the IEEE Standards Association to continue the development of the standard. Currently, a draft standard has been produced but this working group is still in its early stages and as such a final standard is expected to emerge no earlier than late 2007.

1.3 Cellular Networks

Cellular technology has long evolved from first-generation analog technology to today's Internet-enabled digital cellular packet networks. Originally such networks were designed to provide voice service, but today's information-centric users demand other services as well, such as e-mail, text messaging, and wireless Internet browsing. Figure 1.4 illustrates the evolution of cellular technology from second generation to third generation. Here, the evolution of the two primary technologies deployed today is shown: code division multiple access (CDMA) and Global System for Mobile communications (GSM). GSM is largely a time division multiple access (TDMA) system.

The third-generation partnership project (3GPP) was established in December 1998 as a collaboration between multiple regional telecommunications standards bodies: the Association of Radio Industries and Business

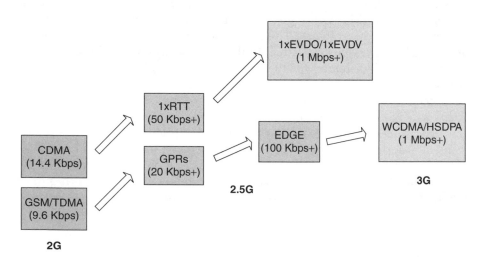

Cellular Technology Evolution

FIGURE 1.4
Evolution of cellular technology.

(ARIB) in Japan, the Telecommunication Technology Committee (TTC) in Japan, the Alliance for Telecommunications Industry Solutions (ATIS) in the United States, the China Communications Standards Association (CCSA) in China, the Telecommunications Technology Association (TTA) in Korea, and the European Telecommunications Standards Institute (ETSI). Together, these standards bodies comprise the organizational partners for 3GPP. The 3GPP project agreement signed by all the organizational partners states that they shall cooperate in producing "globally applicable" technical specifications and reports for a 3G mobile system based primarily on GSM core networks and the radio access technologies they support, such as enhanced data rates for GSM evolution (EDGE), high-speed data packet access, or universal terrestrial radio access (UTRA). The 3GPP was established primarily for preparation, approval, and maintenance of technical specifications and reports for 3G networks based on the GSM core structure. Furthermore, 3GPP is not considered a legal entity.

The 3GPP2 was established in December 1998 as a collaboration between multiple regional telecommunications standards bodies: the ARIB in Japan, the CCSA in China, the Telecommunications Industry Association (TIA) in North America, the TTA in Korea, and the TTC in Japan. Together, these standards bodies comprise the organizational partners for 3GPP2. Also, market representation partners include the CDMA Development Group, the IPv6 Forum, and the International 450 Association. These market representation partners offer market advice and a consensus view on market requirements. The 3GPP2 was established primarily for preparation, approval, and maintenance of technical specifications and reports for 3G networks based on the cdma2000 core network structure. Like 3GPP, 3GPP2 is not considered a legal entity.

The 3G cellular standards addressed by the 3GPP and 3GPP2 can be placed in one of the two respective categories: TDMA or CDMA. TDMA technology operates on the premise that a user on the network has a time slot allocated on the cellular channel. Here, a user occupies the entire bandwidth of that channel for a specified periodic time frame with some period T. Within the length of the period T, many users can occupy the entire bandwidth, as long as each one's time frame does not overlap with the other. As a consequence, accurate, precise timing in a TDMA system from the BS and user perspective is critical. Generally, the bandwidth of each channel is around 200 kHz for GSM-TDMA systems employed today. Furthermore, each channel can hold approximately five to six users at one time. Once all time slots are filled, the TDMA channel is considered to be at full capacity, and no more users can be accommodated until one of the current users disconnects from the system. The advantage of TDMA is that the sound quality is consistent as long as a time slot is available to serve a mobile user. However, once all time slots are filled with mobile users, service is denied to all the other users.

CDMA technology operates quite differently from TDMA. Each user data channel is multiplied by a unique, mathematically orthogonal binary chipping sequence at a much faster rate than the symbol rate of the modulation used. This, in effect, spreads the spectrum of each user to cover a bandwidth of about 1 MHz, so all users share the entire spectrum at the same time and with the same power. Interference is minimized in this approach for two reasons. First, each unique chipping sequence is orthogonal to the next one in signal space. These chipping sequences are called Walsh codes. There are 64 unique Walsh codes. Second, a high-fidelity, rapidly adapting power control mechanism employed at the BS' and mobile users maintain near-equal received power levels from mobile users, as seen by the BS, so no one user has a power advantage over another. Open- and closed-loop power control methods are employed here. The open-loop power control method employs BS observations of power measurements from mobile users. The BS may command a mobile user to adjust its power to match the received signal levels of the other mobile users. The open-loop method operates at a relatively slow rate as compared to closed-loop power control in which the mobile user is an active part of the power control and adjusts its own power based on its observations of received power levels from the BS. CDMA has an advantage over TDMA when considering capacity degradation. While TDMA hard limits the number of users who may use the channel at one time, CDMA allows for a more gradual degradation in quality for each additional user. All active users suffer slight quality degradation when another user joins the network at the same time. However, this can result in a significant variation in sound quality, as compared to the relative consistency of the time slot method employed in TDMA.

While the second generation of both these technologies supported data rates up to 14.4 Kbps (CDMA) and 9.6 Kbps (TDMA), these speeds would not provide the necessary bandwidth to support the applications used on today's wireless Internet architecture. However, evolution to third-generation technology data rates (around a megabit per second) has improved the performance of these high-bandwidth applications.

Today, users have an option with most cellular companies to purchase a Personal Computer Memory Card International Association (PCMCIA) network access card to connect to the Internet. Typical data rates experienced by users range from 300 Kbps up to 1 Mbps, depending on the technology. To date, evolved CDMA technologies such as 1xEVDO have outperformed evolved GSM technologies such as the Universal Mobile Telecommunications System-Wideband CDMA (UMTS-WCDMA) from a data rate perspective. 1xEVDO currently supports a downlink physical layer data rate at 2.4 Mbps and an uplink physical layer data rate at 150 Kbps. Revision A to this standard will improve the downlink physical layer data rate to 3.1 Mbps and increase the uplink physical layer data rate to 1.8 Mbps. The high-speed downlink packet access (HSDPA) standard for UMTS-WCDMA aims to support downlink physical layer data rates from 1.8 up to 7.2 Mbps and beyond by introducing another channel known as the high-speed downlink shared channel (HSDSCH) used solely for downlink communications to the mobile user. The uplink data rate supported by HSDPA is 384 Kbps. More information on these standards and their evolution is discussed in Ref. 5.

Cellular network providers have adopted strategies to evolve their networks to third generation, and most have currently adopted the new technologies available. However, cellular networks are most useful for providing their first envisioned application: voice. Nevertheless, these network providers have noticed the evolving wireless Internet architecture unfold, especially with the success of IEEE 802.11, and as such desire to participate by providing increased data rates and services to compel users seeking wireless network access to utilize the cellular infrastructure. While coverage for cellular networks is by far the most extensive of any wireless network infrastructure deployed to date (with the exception of low-bandwidth satellite), data rates have yet to evolve to support the increasing bandwidth needs of users.

1.4 ETSI HIPERLAN Standard

The ETSI has developed analogous standards to the IEEE 802.11 and IEEE 802.16 solutions, known as HIPERLAN and HIPERMAN, respectively. These technologies are considered to be superior from a throughput and design perspective compared to their IEEE 802 counterparts but nevertheless have not been adopted or deployed widely. In addition to the IEEE 802.16 standard, the WiMAX Forum has also supported certification of ETSI HIPERMAN standards-based equipment.

The HIPERLAN WLAN technology standard [6] was established by ETSI as a way to enable wireless network connectivity on a variety of platforms: third-generation cellular, home wireless LAN, and corporate wireless LAN, for example. The ETSI Broadband Radio Access Networks (BRAN) group has developed the second-generation HIPERLAN/2 as the follow-on standard to HIPERLAN/1, similar to the IEEE 802.11 evolution of standards (from the 11 Mbps "b" standard to the 54 Mbps "g" standard). HIPERLAN/2 operates in the 5 GHz UNII band. It supports data rates ranging from 6 to 54 Mbps via an OFDM format. HIPERLAN/2 uses a TDMA scheme to share the medium among multiple users. Figure 1.5 illustrates the basic architecture of HIPERLAN.

HIPERLAN topologies are similar to cellular infrastructure topologies, in that there are base stations and wireless users. As it complies with the BRAN PHY and data link control (DLC) standards it is interoperable with a variety of other European core network standards such as GSM. There are two modes of operation within HIPERLAN: centralized and direct. Centralized mode is analogous to the infrastructural mode within the IEEE 802.11 standard, where cellular-like infrastructure is required to relay packets from users through base stations to other users. The direct mode is analogous to the *ad hoc* mode in IEEE 802.11, where users can send and receive packets to and from each other without traversing an infrastructure node. HIPERLAN supports

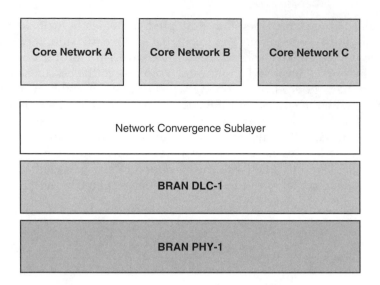

FIGURE 1.5
HIPERLAN architecture.

data rates from 6 up to 54 Mbps, with nominal ranges operating from 100 to 300 m.

There are four OFDM subcarrier modulation modes in HIPERLAN/2: BPSK, QPSK, 16-QAM, and 64-QAM. Mandatory error correction code specifications call for rate 1/2, constraint length $k = 7$ convolutional code, with optional rate 9/16 and 3/4 codes for the higher data rates (27–54 Mbps).

One distinguishing feature of HIPERLAN compared to IEEE 802.11 is that it supports multiple-beam antennas (sectoring) for improved link budget performance and reduction in interference. This feature was included primarily for ease of integration into existing cellular infrastructure. Like IEEE 802.11, however, HIPERLAN increases or decreases data rate by changing modulation and coding based on PHY and MAC layer metrics (such as signal strength and packet loss ratio).

1.5   Bluetooth

The Bluetooth standard [7] was ratified by an industry consortium initially in 1999 to enable short-range wireless connectivity between devices such as PDAs, cellular phones, printers, and computer peripherals. It operates in the 2.4 GHz ISM band with an frequency hopping spread spectrum (FHSS) waveform and has a 400 Kbps data rate (symmetric) or 700 Kbps data rate (asymmetric). The range is about 10 m with a transmitter power of about 1 mW.

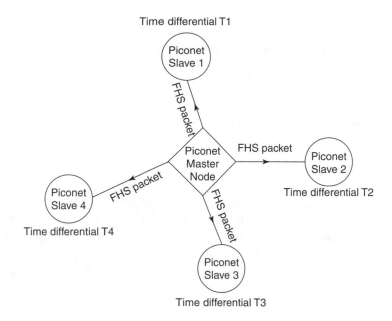

FIGURE 1.6
Bluetooth piconet hierarchy.

Bluetooth forms piconets or associations between nodes based on a particular hopping sequence. Within a piconet there is only one master node. Its clock and Bluetooth device address (BD_ADDR) are passed to slaves via frequency hop synchronization packets. The master BD_ADDR is used to calculate the sequence of frequency hops required for all devices within the piconet to follow to communicate. The master's clock is used to decide which hop in the sequence is current (known as the hopping phase). All slave devices within the piconet use the differential between the master clock and their own to determine which frequency to use at any given time so they can follow the hopping sequence accurately. Each piconet operates on a unique frequency-hopping sequence within the ISM band. Figure 1.6 illustrates a basic hierarchy of a piconet.

Physical channels in Bluetooth are characterized by a single radio frequency combined with temporal parameters and are restricted spatially. Two physical channel types are used for communication between Bluetooth devices: the basic piconet channel and the adapted piconet channel. The other physical channels defined within Bluetooth are used for device discovery within the Bluetooth domain (inquiry scan channel) and for establishing connection between Bluetooth devices (page scan channel).

While the Bluetooth standard has been adopted as an IEEE 802 standard (as IEEE 802.15.1), it was defined prior to standardization by IEEE 802 and has been deployed significantly as a feature of wireless cellular handsets and handset accessories such as headsets.

1.6 Other Wireless Networking Technologies

This chapter has delineated and briefly described some of the wireless networking technologies that are expected to compete with or compliment WiMAX. Other technologies will most certainly emerge and as such may change the marketplace climate for WiMAX and related technologies significantly. Such technologies include

- The Wireless Broadband (WiBro) standard, a Korean standard that is incorporated into the IEEE 802.16e standard. While the WiMAX Forum has indeed created certification profiles for IEEE 802.16, it is unclear as to what degree the WiMAX Forum will certify WiBro-enabled equipment. This could affect deployment of the WiBro technology.
- The IEEE 802.22 wireless regional area network (WRAN) standard is currently emerging and aims to employ cognitive radio concepts to enable a next-generation adaptive wireless networking technology operating in the licensed broadcast television bands.

1.7 Competing Technologies

Some of the wireless technologies described here will undoubtedly compete with WiMAX and its associated technologies. This section of the chapter provides a discussion of such technologies and what advantages or disadvantages each has when compared to WiMAX.

1.7.1 IEEE 802.20

The IEEE 802.20 working group and IEEE 802.16 Task Group E have been widely considered as developers of competing technologies. However, there are some differences between the two standards:

- IEEE 802.20 aims to develop a standard that supports 1 Mbps data rates for mobile users moving at speeds up to 250 km/h. IEEE 802.16e, however, only supports users at vehicular speeds, notionally up to 150 km/h.
- IEEE 802.16e is intended for frequencies operating from 2–6 GHz. However, IEEE 802.20 is focused on frequencies at 3.5 GHz or below.
- IEEE 802.16e is based on prior IEEE 802.16 standards work, while IEEE 802.20 aims to produce an original standard.

- IEEE 802.16e is a ratified standard while IEEE 802.20 is still in the draft form. Furthermore, contention in the IEEE 802.20 working group may prevent any final version of the standard, similar to what has happened in the IEEE 802.15.3a working group.

It is expected that IEEE 802.20 will not support the high data rates that IEEE 802.16 provides, as the solution space for IEEE 802.20 focuses on high-speed mobility. However, as the IEEE 802.16e standard evolves and WiMAX profiles are defined for various mobility classes, advancements in technology and methods could improve mobility support up to and surpassing speeds defined in IEEE 802.20 for implementation-specific IEEE 802.16e equipment. Furthermore, a wide variety of industry participants have embraced IEEE 802.16 and WiMAX certification as the path to broadband wireless mobile access, although IEEE 802.20 aimed to produce a standard that achieved this vision.

1.7.2 Cellular Networks

Perhaps the largest competitor to WiMAX technologies, cellular networks have been deployed all over the world. The level of investment and infrastructure deployment has been one of the most extensive of any terrestrial wireless network in existence. Furthermore, paths to evolve to higher data rates that support mobility from the start and provide users with an experience that approximates the wired connectivity they experience at home or at the office is expected to materialize as the technologies evolve. However, there are some disadvantages to cellular as compared to WiMAX:

- *Cost*: The expense of procuring and deploying a cellular network infrastructure with the most advanced, high-data-rate technologies (such as HSDPA and 1xEVDO) today is substantially larger than a WiMAX-enabled solution. First, the cost of maintaining spectrum licenses for cellular bands is substantial. Furthermore, base station cost is about an order of magnitude more expensive to procure. Finally, the complexity of such a solution is significant especially when cellular providers must retrofit their older-generation networks and maintain separate networks to ensure users without the latest equipment will be able to maintain access.
- *Original design*: Cellular systems were originally designed for voice communications and as such have been augmented to support a variety of data applications. The CDMA and GSM core networks have also evolved to support IP-based communications, which has become the *de facto* standard today. However, WiMAX technologies are primarily IP-based and were designed to support data and voice applications from the beginning.
- *Throughput performance*: Results in Ref. 8 suggest that, when all other system parameters remain equal (bandwidth, antenna

configuration, power), WiMAX technology outperforms both HSDPA and cdma2000 3xEVDO (three 1xEVDO channels) by 28%–96%.

While there are some clear disadvantages to cellular, its key advantage over WiMAX technology remains its large coverage footprint. However, some cellular service providers such as Sprint Corporation in the United States have announced plans to deploy WiMAX across their entire coverage footprint as well. In this sense, WiMAX would be considered complementary to cellular. Many other carriers, especially those with heavy investments in the GSM/UMTS-WCDMA technology space, have not adopted the same coexistence strategy. Sprint's success or failure in deploying WiMAX on a nationwide scale will likely affect similar companies' strategies in dealing with WiMAX competition and deployment.

While the technologies presented in this section are expected to compete for users that WiMAX aims to serve currently, market forces could very well align these technologies with WiMAX in a strategy to further the deployment and use of the emerging wireless Internet architecture.

1.8 Complementary Technologies

Technologies presented in this section are largely complementary to WiMAX. These technologies have been proven for their intended purposes and do not overlap significantly compared to the purposes WiMAX technologies were designed to serve.

1.8.1 IEEE 802.11

The widespread adoption of IEEE 802.11 has resulted in a substantial increase in the ability to connect to the Internet wirelessly. However, unlike WiMAX, IEEE 802.11 was primarily designed for local area networks. It lacks the complexity and power levels inherent in WiMAX that would be required for scalability while maintaining high levels of throughput. WiMAX technologies are primarily based on time division duplex (TDD) or frequency division duplex (FDD) access methods with access slots reallocated to users as needed on a demand basis. IEEE 802.11, however, shares the media with multiple users by employing a distributed-approach version of Carrier Sense Multiple Access with Collision Avoidance (CSMA/CA), which is inherently limited when trying to scale one single channel to support many users. In this sense, the IEEE 802.11 WLAN standards address connecting local areas (within 10s or 100s of feet) most efficiently—one would not prefer the complexity of a WiMAX base station to support connections with maximum distances in this range. It is expected that within short ranges and limited number of users, IEEE 802.11 will outperform WiMAX technologies significantly. WiMAX deployments will not possess the bandwidth required to support many users in a city-wide coverage area with the data rates each would experience if there were one IEEE 802.11 access point for every few users. Such a scenario is typical of the home-network model, where one IEEE 802.11 access point is deployed, connected to a wired infrastructure such as a cable modem or digital subscriber line (DSL).

1.8.2   IEEE 802.15

The IEEE 802.15 family of standards focuses primarily on WPANs with ranges only on the order of 10 ft. Obviously, WiMAX technologies were not designed with this limited range in mind, but the need and demand for WPANs have become increasingly prevalent as wireless networking evolves to support a variety of platforms, including those in the home such as household appliances. Furthermore, mobile phones enabled with IEEE 802.15 technologies benefit from the ability to connect to other phones, computers, or devices such as headsets, albeit within a short range. As the data rate requirements for the applications running over WPANs remain relatively small compared to WiMAX technologies, this technology clearly has delineated a niche compared to the intended use for WiMAX.

1.9   Conclusion

The momentum built up behind WiMAX technologies has reached a critical point. Significant investment in research, development, products, and marketing for WiMAX has been ongoing and is expected to continue. This chapter has described other wireless networking technologies that compliment or compete with WiMAX technologies. It has also provided an overview of the most prevalent technologies in use today, as well as a description of the similarities and differences compared to WiMAX.

References

1. IEEE 802.11-1999, *Part 11: Wireless LAN Medium Access Control (MAC) and Physical Layer (PHY) Specifications*, 1999.
2. IEEE 802.11a-1999, *Part 11: Wireless LAN Medium Access Control (MAC) and Physical Layer (PHY) Specifications: High-Speed Physical Layer in the 5 GHz Band*, 1999.
3. IEEE 802.11b-1999, *Part 11: Wireless LAN Medium Access Control (MAC) and Physical Layer (PHY) Specifications: Higher-Speed Physical Layer Extension in the 2.4 GHz Band*, 1999.

4. IEEE 802.11g-2003, *Part 11: Wireless LAN Medium Access Control (MAC) and Physical Layer (PHY) Specifications: High-Speed Physical Layer in the 2.4 GHz Band*, 2003.
5. IMT-2000 Network Aspects, http://www.itu.int/ITU-T/imt-2000/network.html.
6. HIPERLAN 2 Specification, http://portal.etsi.org/bran/kta/Hiperlan/hiperlan2tech.asp.
7. *Specification of the Bluetooth System: Wireless Connections Made Easy*, Version 1.2, 5 November 2003.
8. *Mobile WiMAX—Part II: A Comparative Analysis*, WiMAX Forum, Copyright 2006.

2

IEEE 802.16 Standards and Amendments

Najah Abu Ali and Hossam S. Hassanein

CONTENTS

2.1 Introduction

The IEEE Standards Board established the IEEE 802.16 working group in 1999 to prepare formal specifications for global deployment of broadband wireless metropolitan area networks, which is officially called WirelessMAN. The IEEE 802.16 working group, which is a unit of the IEEE 802 LAN/MAN Standards Committee, is responsible for framing specifications of the IEEE 802.16 family standard, but not testing them. Thus, another industrial group was established in April 2001 called the WiMAX Forum. The acronym WiMAX expands to "Worldwide Interoperability for Microwave Access." WiMAX Forum is on a mission to advance and certify compatibility and interoperability of broadband wireless products based on IEEE 802.16 family standards. Irrespective of the scope of the WiMAX Forum that aims to test equipments, the IEEE 802.16 family hails WiMAX from the WiMAX Forum, maybe because it is easier to use the word WiMAX rather than IEEE 802.16.

2.2 Standardized versus Proprietary Solutions

Before proceeding to present the developments of the 802.16 family of standards, it is worthwhile to know the pros and cons of the standardized versus proprietary solutions in WiMAX case (Alvrion, 2005).

2.2.1 Standardization Cons

1. Setting rules normally consumes long periods of time before being available to vendors. This may encourage a change to another technology that provides the same service, for example, 3G.
2. Gaining agreement across the standards committee may require degrading the specifications to gain the common players' approval. Consequently, the resulting standard may not satisfy the user or at least the counterpart proprietary solution may provide a superior technical performance.
3. Forcing the vendors to comply with a standard may hinder vendors from competition to produce innovative solutions.

2.2.2 Standardization Pros

1. Reduces supplier dependence resulting in a wider deployment of the technology because there is no dependence on a sole producer
2. Lowers the product cost and consequently lowers the cost to the end user
3. Lowers the deployment risk owing to interoperability

However, by comparing the IEEE 802.16 family of standards with other existing standards, we can see that the standardization process did not extend over a considerably long time. Additionally, the IEEE 802.16 family of standards includes a wide range of variations (as we will see in the following sections). Hence, while being standard compliant, it leaves breathing space for solution innovation by vendors.

2.3 Overview of the Standard

IEEE 802.16-2001, the first standard of the family, was approved in December 2001 and published in 2002. This standard is the result of the activity of hundreds of participants worldwide. The working group of this standard (Air Interface for Fixed Broadband Wireless Access System) focused on providing WirelessMAN access for fixed applications. IEEE 802.16-2001 (LAN/MAN committee, 2001) provides network access to buildings through exterior antennas communicating with a radio base station using point-to-multipoint (PMP) infrastructure design and operating at a radio frequency between 10 and 66 GHz with an average bandwidth performance of 70 Mbps and a peak rate up to 268 Mbps. Thus, it is basically an alternative to cabled access networks, cable modems, and digital subscriber line (DSL). However, the IEEE 802.16-2001 standard was not an adequate air interface standard for broadband wireless access. It addressed frequencies in a licensed spectrum that introduces significant challenges to the short wavelength and is limited to line-of-sight (LOS) propagation. It also neglects any conformance with its European counterpart standard, HiperMAN standard, and supports a single-carrier physical layer. Thus, the initial 802.16-2001 standard was followed by several amendments.

The first one was IEEE 802.16c (LAN/MAN committee, 2002). The main objective of this amendment was to ensure interoperability among the existing local multipoint distribution service (LMDS) LOS solutions working in the 10–66 GHz range. Naturally, since the 802.16c is defined over a wide range of frequency it provides more bandwidth. However, and for the same reason, the maximum coverage of 802.16c does not exceed 5 km. In addition to 802.16c's main objective, it addressed other issues such as testing, performance evaluation, and system profiling. System profiling is a vital requirement for interoperability. 802.16c provides guidelines for vendors through mandatory and optional elements of system profiling to ensure interoperability. As for mandatory elements of 802.16c profiling, vendors should support provisioned connections, provide IPv4 support on transport connection, and support fragmentation. As for optional elements, 802.16c allows for different levels of security protocols that allow vendors to provide different functionalities that differentiate their products. As a final remark on 802.16c, it is specified to be network technology independent. Thus it can run under asynchronous transfer mode (ATM), internet protocol (IP), or frame relay. The second amendment was the IEEE 802.16b, also called WirelessHUMAN (Wireless high-speed unlicensed metropolitan area network). This amendment mainly provided for quality of service (QoS) features to ensure differentiated service levels for different traffic types. It extended 802.16-2001 to operate under license-exempt regulation in the 5–6 GHz range. However, 802.16b does not exist anymore. In April 2003, 802.16a, the most eminent among amendments, was published to standardize the lower-frequency multichannel multipoint distribution service (MMDS) solutions in the licensed and unlicensed range of 2–11 GHz. Working at a lower-frequency range than 802.16-2001, 802.16a (LAN/MAN committee, 2003) has the advantage of being able to offer nonline-of-sight (NLOS) communication and a cell coverage up to 50 km with a bit rate up to 75 Mbps. An additional feature of 802.16a is that it provides for mesh mode operation, which facilitates subscriber-to-subscriber communications. IEEE 802.16d project was launched to produce interoperability specification and to provide for some fixes for 802.16a. However, the project was transitioned into a revision project for 802.16-2001 and all its amendments. The revision project result is no longer called 802.16d, but it is formally called 802.16-2004 (LAN/MAN committee, 2004). Yet, this active standard was followed by different working groups to address different issues as follows:

1. Active standards
   a. IEEE 802.16e-2005 (formerly known as IEEE 802.16e)—addressing mobility, concluded in 2005
   b. 802.16f—Management Information Base
2. Drafts under development
   a. 802.16g—Management Plane Procedures and Services
   b. 802.16k—Bridging
   c. 802.16h—Improved Coexistence Mechanisms for License-Exempt Operation
3. Projects in predraft stage
   a. 802.16i—Mobile Management Information Base
   b. 802.16j—Mobile Multihop Relay

In the following sections, we present the IEEE 802.16-2004 standard and its amendments, their status, and an overview of their specifications.

2.4 IEEE 802.16-2004

As aforementioned, the first standard of 802.16 addressed the LOS communication in the 10–66 GHz band. 802.16a extended its operation to include

NLOS communication in the lower-frequency band of 2–11 GHz. Thus, IEEE 802.16-2004 (LAN/MAN committee, 2004) supports communication in the 2–66 GHz band. LOS and NLOS propagation are quite different. Thus, to design a standard that supports both bands, the physical and the medium access control (MAC) layer should support these differences. For example, signal propagation in high-band frequency is highly affected by obstacles, consequently LOS propagation is utilized, which in turn results in alleviating the effect of multipath interference. Multipath results from receiving the signal at the receiver through more than one path owing to reflection and refraction of obstacles. However, operation in the lower band that includes licensed and unlicensed spectrum requires its own regulations. For example, operation in the unlicensed spectrum requires management of transmitter output power, techniques to avoid frequency interference, etc. These issues and others not only affect the physical layer design but also influenced the MAC layer. Thus, the scope of 802.16-2004 standard covers the specifications of these two lower layers in the OSI model.

2.4.1 Physical Layer

A 10–66 GHz frequency wave is a focused beam, which theoretically can reach multiple miles through LOS propagation. Designers deemed that single-carrier modulation was a sufficient choice and the physical layer standard version of this band is called WirelessMAN-SC (single carrier). WirelessMAN-SC can support frequency division duplex (FDD) and time division duplex (TDD) modes. However, operation in the 2–11 GHz band required changes in the physical layer specification to support NLOS propagation. Mainly, three new PHYsical layer (PHY) specifications were introduced to meet this requirement—a single-carrier PHY, a 256-point FFT OFDM PHY, and a 2048-point FFT OFDMA PHY. The single-carrier PHY, designated as WirelessMAN-SCa, is based on the WirelessMAN-SC. However, there are some differences such as framing elements that enable improved equalization and channel estimation performance over NLOS propagation, extended delay spread channels, parameter settings, and MAC/PHY messages that facilitate optional adaptive antenna systems (AAS) implementations. The second and third PHY specifications employ orthogonal frequency division multiplexing (OFDM), which is a multicarrier transmission technique suitable for high-speed NLOS. OFDM uses 256 RF subcarriers to transmit different signals simultaneously. The neighboring subcarriers are allowed to overlap; however, they are orthogonal to each other to prevent inter-carrier interference (ICI). The key difference between WirelessMAN SCx and OFDM is that OFDM is more resilient to the multipath effect. OFDM has higher bandwidth efficiency since it allows neighboring subcarriers to overlap. Thus, OFDM modulates data at a rate of 72 Mbps over a channel bandwidth of 20 MHz, which provides a spectral efficiency of 3.6 bps/Hz (WiMAX Forum, 2004).

Orthogonal frequency division multiple access (OFDMA) is a 2048 sub-carrier OFDM scheme. The difference between OFDM and OFDMA is that OFDMA organizes the time (i.e., the symbols) and the frequency (i.e., sub-carriers) resources into subchannels for allocation to individual receivers, which allows for multiple access. Thus, OFDMA operates over two dimensions, time and frequency. There are two types of subcarrier permutations for subchannelization—diversity and contiguous (WiMAX Forum, 2006). The diversity permutation draws subcarriers pseudorandomly to form a subchannel. The contiguous permutation groups a block of contiguous subcarriers to form a subchannel. OFDM PHY is common between 802.16 and ETSI HiperMAN because, for example, OFDM requires weaker frequency synchronization and faster Fast Fourier Transform (FFT) calculation. Consequently, WiMAX Forum focuses on 256-carrier OFDM PHY in all its profiles.

One may ask, why not use code division multiple access (CDMA) as a signaling format? CDMA requires a bandwidth that is much larger than the data throughput to maintain a processing gain capable of overcoming interference. Furthermore, OFDM and OFDMA support NLOS performance making maximum use of the available spectrum.

2.4.1.1   Other Features

The PHY layer also has other features, some of them are mandatory and the others are optional. These features empower the performance of the technology to provide for robust performance over a wide range of frequencies and under different channel conditions.

- Adaptive antenna system (AAS): uses multiple antennas at both the receiver and the transmitter ends (MIMO system) to increase channel capacity by steering the antenna beams toward multiple users to achieve in-cell frequency reuse. MIMO system is also beneficial in increasing the signal-to-interference ratio through coherently combining multiple signals. Another benefit of AAS is the decrement of required power due to utilizing beams formed of adaptive antennas.

- Adaptive modulation: 802.16-2004 allows for different modulation schemes in the down- and uplink communication, i.e., BPSK, QPSK, 16QAM, 64QAM, and 256QAM. The 802.16 standard defines different combinations of the aforementioned modulation schemes and coding rates, providing for a wide range of trade-offs of data rate and robustness depending on channel conditions. Although 802.11a/g standard uses similar modulation schemes as 802.16, there is one difference between them, 802.16 uses Reed–Solomon block code with an inner convolution code or Turbo coding. The latter is left as an optional feature.

- Space time coding: is an optional feature of 802.16 that can be used in the downlink communication to provide for space transmit diversity. Space time coding assumes that the base station is using two transmit antennas and the subscriber station uses one transmit antenna.

2.4.2 MAC Layer

The MAC layer supports the different PHY specifications by using time division multiplexing, where users are assigned time slots to access the channel. The uplink communication is based on time division multiple access (TDMA). TDMA facilitates different levels of QoS and bounded delay communication through a predetermined service level agreement. This can be achieved by allocating bandwidth based on a request/grant mechanism. The standard 802.16-2004 supports both TDD and FDD, full and half duplex.

802.16-2004 is designed to carry any present or future higher-layer protocol such as IP versions 4 and 6, packetized voice-over-IP (VoIP), Ethernet, ATM, and virtual LAN (VLAN) services. 802.16 accomplishes this by dividing its MAC layer into separate sublayers that handle different services as follows:

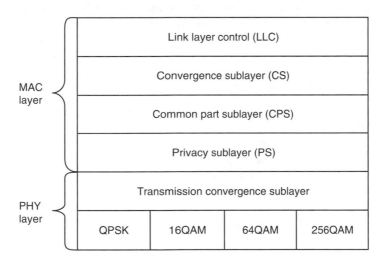

2.4.2.1 *Convergence Sublayer*

Convergence sublayer (MESA, 2005) is designed to map services to and from 802.16 MAC. 802.16 has two services—the ATM convergence sublayer and packet convergence sublayer. Packet convergence sublayer provides support for IPv4, IPv6, Ethernet, and VLAN. The main task of the convergence sublayer is to map higher protocol data units into proper service data units. Additionally, it is responsible for bandwidth allocation and QoS, as well as header suppression and reconstruction to enhance air-link efficiency.

2.4.2.2 *Common Part Sublayer*

802.16 is designed to support PMP network architecture. However, mesh operation, also known as point-to-point architecture, is left as an optional feature.

2.4.2.2.1   802.16 PMP

IEEE 802.16 MAC protocol was designed as PMP connection from the base station (BS) with sectorized antenna to multiple subscriber stations (SSs). TDD multiplexing is used to divide transmission time into up- and downlink periods. On downlink, data to SSs are multiplexed in TDM fashion and generally broadcasted to all SSs capable to listen to the downlink frame. Each SS checks the connection ID in the protocol data units (PDUs) and retains the PDUs addressed to it. The uplink is shared between SSs implementing TDMA on demand bases.

802.16 MAC is connection oriented even for connectionless transmissions such as IP. Connectionless transmission is mapped into a connection, which is used as a pointer to destination and context information. In the 802.16 standard, SSs are identified by a 48-bit universal MAC address. This address is unique and is normally used for authentication and during ranging process to establish connections. Connections are referenced with 16-bit connection identifiers (CID). Upon joining the network, three management connections and at least one transport connection are established between the BS and SS in the up- and downlink direction. The third management connection is optional. The three management connections reflect different levels of QoS as follows:

1. Basic connection: short-time urgent MAC management messages
2. Primary management connection: longer, more delay-tolerant messages
3. Secondary management connection: standard-based delay tolerant management messages such as Dynamic Host Configuration Protocol (DHCP), Trivial File Transfer Protocol (TFTP)- and Simple Network Management Protocol (SNMP)

Transport connections are used to facilitate different QoS communication levels for the up- and downlink. The contracted level services are unidirectional, thus, the QoS level may differ between the up- and downlink. In addition to the aforementioned connections, there are three additional special purpose connections. One is reserved for connection-based initial access while the other two are broadcast and multicast connection based polling.

802.16 defines the concept of service flow. Once an SS joins a network and connection is established, the connection is mapped into the service flows; each connection is mapped to one service flow. Service flows provide a mechanism for up- and downlink QoS management, mainly the bandwidth allocation process. Bandwidth is allocated to an SS by a BS as a response to a per connection request from the SS. Bandwidth allocation may be constant depending on the type of service, for example, T1 unchannelized services or it may be adaptive such as that granted for the IP bursty services. As with connection establishment, connections may undergo maintenance or termination.

2.4.2.2.2   802.16 Mesh

The key difference between the PMP and mesh topology is that in the PMP mode communication is based on a direct connection between the BS and SSs, while in the mesh mode multihop communication is allowed, where traffic can be routed through other SSs and can occur directly between SSs. Hence, an SS may operate as a router to relay traffic between SSs until it arrives to a BS, called mesh BS. Mesh BS has a direct connection to backhaul services outside the mesh network. All the other systems of a mesh network are termed mesh SS nodes. In mesh node, the term up- and downlink has a different meaning. Uplink is defined as traffic in the direction of the mesh BS while downlink is defined as traffic away from the mesh BS.

In mesh mode the up- and downlink transmission is still based on TDMA. However, mesh mode defines another type of scheduling mechanism in addition to the centralized scheduling, distributed scheduling and a combination of both distributed and centralized scheduling. In centralized scheduling, a mesh BS gathers resource requests from all the mesh SSs within a certain hop range. It determines the amount of granted resources for each link in the network, both in down- and uplink, and communicates these grants to all the mesh SSs within one hop range (LAN/MAN committee, 2004). In distributed scheduling, all nodes including the mesh BS coordinate their transmissions in their two-hop neighborhood and broadcast their available resources, requests, and grants to all their neighbors. 802.16 defines a node neighbor to be nodes one hop away (forming nodal neighborhood). Additionally, distributed scheduling can be established by directed uncoordinated requests and grants between two nodes. Hence, communicating nodes are required to ensure collision-free transmission within two hops proximity.

We remark that the 802.16 mesh operates in the licensed and unlicensed 2–11 GHz NLOS communication spectrum. We also remark that QoS over the mesh is link based; there is no end-to-end QoS guarantees. QoS is provisioned over links on a message-by-message basis, where each message has service parameters in its header.

2.4.2.3   Privacy Sublayer

Privacy sublayer is a separate security sublayer that provides secure key exchange and encryption. Privacy sublayer has two main protocols

1. An encapsulation for encrypting packet data across the 802.16 network.
2. A privacy key management (PKM) protocol to facilitate secure distribution of the keying data from the BS to the SS. PKM is enhanced by adding digital-certified-based SS authentication to be used in the 802.16 architecture.

PKM is used in security association, which is a set of cryptographic methods and the associated keying material. 802.16 defines three types of security association—primary, static, and dynamic. Primary security association is established during the SS initialization. Static service association is provisioned within the BS while dynamic security association is initiated and terminated on demand in a response of initiation and termination of service flows.

2.5   IEEE 802.16e-2005

On July 2002, a study group called IEEE 802.16 Mobile WirelessMAN Task Group was initiated to produce an amendment covering the PHY and MAC layers for combined, fixed, and mobile operations in the licensed band range. The amendment was approved in December 2005 and the new standard called IEEE 802.16e-2005 was published in February 2006. The scope of this standard is to provide mobility enhancement support for SS moving at the vehicular speed, in addition to corrections to 802.16-2004 fixed operation that was developed as IEEE 802.16-2004/Cor1-2005 and published along with IEEE 802.16e-2005. 802.16e (IEEE 802.16e-TG, 2006) introduces many changes to PHY and MAC layer protocols owing to mobility support, which required addressing new issues that were not required in 802.16-2004, such as handoff and power management.

2.5.1   Physical Layer

IEEE 802.16e-2005 is an amendment to IEEE 802.16-2004. Thus, we restrict our discussion to the changes to the PHY layer introduced by IEEE 802.16e-2005:

1. 802.16e operation is limited to licensed bands suitable for mobility below 6 GHz. This may introduce a compatibility problem between 802.16-2004 and 802.16e, since the available licensed spectrum may need to be split between the two technologies.
2. 802.16e defines a new PHY air interface, scalable-OFDMA (S-OFDMA), besides those defined by 802.16-2004. S-OFDMA uses FFT size of 128, 512, 1024, or 2048 subcarriers. S-OFDMA uses this number of subcarriers to provide the ability to scale system bandwidth while at the same time the subcarrier separation and symbol duration remain constant as the bandwidth changes. Thus, the BS determines the subcarrier used to adapt to its devices' channel conditions.
3. The AAS, space time code, and closed-loop MIMO modes are enhanced in 802.16e to improve coverage and data transmission rate. Additionally, support for coordinated spatial division multiple access (SDMA) is introduced (Motorola, 2005).
4. 802.16e includes an additional advanced low complexity coding option method, low-density parity check (LDPC) to provide for more flexible encoding. LDPC codes 6 bits for every 5 data bits with

a rate of 5/6. This forces higher-performance coding technique than the methods included in 802.16-2004 that provide 3/4 code rate.

2.5.2 MAC Layer

MAC layer specification practices considerable departures from 802.16-2004 to provide support for mobility. It adds support for handoff and power management.

2.5.2.1 QoS Support

802.16e defines new scheduling mechanisms: the extended real-time polling service (ErtPS), which is based on two services defined in 802.16-2004; the unsolicited grant service (UGS); and the real-time polling service (rtPS). ErtPS is similar to UGS in providing unicast grants, thus saving the delay incurred for requesting the bandwidth. However, ErtPS allocations are dynamic as rtPS while UGS allocations are fixed. The ErtPS is introduced to support real-time service flows that generate periodical variable sized data packets. Thus, ErtPS is especially important to support VoIP, since it allows for managing traffic rates and improves latency and jitter.

2.5.2.2 Handover Support

802.16e includes new MAC-level request/grant mechanisms to achieve similar seamless mobility as that provided for cellular users. 802.16e includes fast base station switching and hard handoff mechanisms for intercell and intersector handover. In 802.16e, handoff process may be triggered for two reasons. One is due to fading of the signal, interference level, etc. within the current cell or sector. The other is due to the fact that another cell can provide a higher level of QoS for the mobile station (MS). Furthermore, 802.16e supports macrodiversity handovers and intertechnology roaming. Macro-diversity handovers support handoffs between different sized cells, while intertechnology roaming addresses MS handoffs from BS to backhaul or wired network by providing roaming authentication mechanisms.

2.5.2.3 Power Management

Power management is a critical process for mobile applications to enable efficient operation of the MS. 802.16e defines two power management operations, sleep mode and idle mode.

Idle mode operation is carried out by MS when the MS does not intend to register to a specific BS as the MS traverses a region covered by multiple BS. The advantage of idle mode for the BS is to avoid multiple handoffs and other normal operations while the SS is traversing the region, and for the BS and network is to avoid unnecessary handoffs from an inactive MS. When the MS enters the idle mode, it needs to periodically check for broadcast messages sent by the BS to see if new downlink frames have been sent to it (WiMAX Forum, 2006).

Sleep mode operation is a state in which MS sends a request to be unavailable to the BS. If the BS responds with approval, the MS is provided with a sleep interval time vector that determines the length of the sleep mode period. The benefit of the sleep mode operation is to minimize MS power usage and utilization of the air interface resources of the BS. While the MS is in the sleep mode, the MS scans other BSs to collect information required for handover during the sleep mode.

2.6   IEEE 802.16f

IEEE 802.16's Network Management Study Group was created in August 2004. Its scope of work was to define a management information base (MIB) for the MAC and PHY, and associated management procedures. The working group approved 802.16f amendment that provides MIB for fixed broadband wireless access system in September 2005.

IEEE 802.16f (IEEE NetMan, 2005b) provides a management reference model for 802.16-2004 based networks. The model consists of a network management system (NMS), managed nodes, and service flow database. The BS and managed nodes collect the required management information and provide it to NMSs via management protocols, such as Simple Network Management Protocol (SNMP) over the secondary management connection defined in 802.16-2004. IEEE 802.16f is based on the SNMP version 2 (SNMPv2), which is backward compatible with SNMPv1. 802.16f provides optional support for SNMPv3.

2.7   IEEE 802.16i

IEEE 802.16i project was initiated in December 2005 within the Network Management Study Group to amend or supersede 802.16f. 802.16i is currently in its early phase, the predraft stage. The scope of 802.16i is to provide mobility enhancements to 802.16 MIB to the MAC layer, PHY layer, and associated management procedures. It uses protocol-neutral methodologies for network management to specify resource models and related solution sets for the management of devices in a multivendor 802.16 mobile network (IEEE NetMan, 2006b).

2.8   IEEE 802.16g

IEEE 802.16g (IEEE NetMan, 2005a) project was initiated in August 2004 within the Network Management Study Group. The scope of 802.16g is to produce procedures and service amendments to 802.16-2004 and 802.16e-2005; provide network management schemes to enable interoperable and efficient management of network resources, mobility, spectrum; and standardize management plane behavior in 802.16 fixed and mobile devices.

802.16g defines a generic packet convergence sublayer (GPCS) as upper layer protocol-independent packet convergence sublayer that supports multiple protocols over 802.16 air interface. GPCS was designed to facilitate connection management by passing information from upper layer protocols without a need to decode their headers. This is achieved by allowing the upper layer protocols to explicitly pass information to the GPCS service access point (SAP) and map the information to the proper MAC connection. GPCS provides an optional way to multiplex multiple layer protocol types over the same 802.16 connection. GPCS is not meant to replace any convergence sublayer (CS) defined by other 802.16 standards or amendments.

Given that 802.16 devices may be part of a larger network, they require interfacing with entities for management and control purposes. 802.16g abstracts a network control and management system (NCMS) that interfaces with the BSs. 802.16g is only concerned with the management and control interactions between MAC/PHY/CS layers of the 802.16 devices and the NCMS. NCMS consists of different service entities such as paging services, gateway and router services, network management multimedia session services, interworking services, synchronization services, data cache services, coordination services, management services, security services, network management services, and media-independent handover function services. These entities may be centralized or distributed across the network. The details of the various entities that form the NCMS as well as the protocols of NCMS are kept outside the scope of 802.16g. NCMS handles any necessary inter-BS coordination that allows 802.16 PHY/MAC/CS layers to be independent of the network and thus allow more flexibility on the network side. 802.16g is still under development. It is expected that 802.16g will be submitted for approval by the start of 2007.

2.9 IEEE 802.16k

IEEE 802.16k (IEEE NetMan-TG, 2006a) was created in March 2006 by the Network Management Study Group to develop a series of standards as amendments to IEEE 802.16 and IEEE 802.1D for 802.16 MAC layer bridging. The 802.16k study group is working to define the necessary procedures and MAC layer enhancements to allow 802.16-2004 to support bridge functionality defined in 802.1D. Transparent bridges assume LAN-like communication of all 802.x technologies, where transmission of one node is heard by all nodes on the same LAN. However, 802.16-2004 devices may filter transmission by address, preventing its attached bridges from bridge address learning. 802.16k (Johnston, 2006) addresses this problem by describing how the internal sublayer service (ISS) is mapped onto the 802 convergence sublayer and how the packets are subsequently treated so that the service below the ISS closely models LAN behavior sufficiently so that the bridge can work. Furthermore, 802.16k provides explicit support for 802.1p end-to-end priority data through explicit one-to-one mapping of user priority.

2.10   IEEE 802.16h

IEEE 802.16's License-Exempt (LE) Task Group was initiated in December 2004 to develop a standard to improve coexistence mechanisms for license-exempt spectrum operation. The main purpose of IEEE 802.16h (IEEE LE-TG, 2006) is to develop improved MAC mechanisms to enable coexistence among licensed-exempt 802.16-2004 devices and facilitate coexistence with other systems using the same band. The amendment is in process, with scope for producing mechanisms that are applicable for the whole uncoordinated frequency spectrum defined by 802.16-2004.

802.16h designs a coexistence protocol, which is defined at the IP level and is mainly intended for BS-BS communication. The coexistence protocol introduces mechanisms for rental and negotiation of spectrum radio resources between BSs within the interference range. The procedures used by the coexistence protocol for interference resolution is based on separating the interference in the frequency and time domains. The separation of interference in the frequency domain is undertaken first, followed by the separation of remaining interference in the time domain.

2.10.1   MAC Enhancement for Coexistence

802.16h is in the process of providing MAC enhancements to support communication using license-exempt and uncoordinated bands. We list below some of the enhancements included in (IEEE LE-TG, 2006). A complete description of these enhancements was not ready until the time of writing this document.

1. Capability negotiation: is a mechanism provided at the MAC layer for the BS to learn about its associated SS capabilities and functionalities for supporting coexistence licensed-exempt band.

2. Extended channel numbering structure: is used to define the channel bandwidth for better interference management. This procedure provides enhancement to channelization and definition of channel numbers. It defines three channelization schemes— extended channel number, which specifies channel number reference; base channel reference, which defines the frequency range; and channel spacing, which defines channel spacing value in 10 kHz increments.

3. Measurement and reporting: a process for defining mechanisms and messages at the MAC layer to measure and report interference level and bandwidth band usage.

2.11   IEEE 802.16j

IEEE 802.16's Relay Task Group is in charge of developing amendments to extend the IEEE 802.16e-2005 to support multihop relay operation. IEEE 802.16's Mobile Multihop Relay Study Group was in charge of IEEE 802.16j project since July 2005. The study group was disbanded in March 2006 and the project was assigned to the Relay Task Group, which continues to work on the project that is still in the predraft phase.

802.16j (IEEE Relay-TG, 2006) is intended to improve legacy 802.16 network's coverage, throughput, and system capacity. 802.16j extends the network infrastructure of legacy 802.16 to include three relay types: fixed relays, nomadic relays, and mobile relays. 802.16j is required to enable the operation of the relay nodes over the licensed band. The OFDMA PHY air interface is the PHY layer specification chosen by the group for 802.16j operation. 802.16j is supposed to define the necessary MAC layer enhancements while at the same time it does not change the SS specifications. However, existence of mobile relay types requires that the relaying process should be carried out by the MS as well. To provide an efficient relaying process, MS should be chosen efficiently and should have some knowledge of the network status, mobility characteristics of other MSs, and the traffic. Thus, conventional MS may not serve as a mobile multihop relay (MMR), since relay stations (RS) are required to pretend to be a BS for MS and to be an MS for BS. Hence, 802.16j defined the three RS types capable of supporting PMP links, MMR links, and aggregation of traffic from multiple RSs. To facilitate RSs communication with BS, this requires changes to BS to support MMR links and aggregation of traffic from multiple RSs. To achieve MMR requirements, 802.16j enhances the normal frame structure at the PHY layer and adds new messages for relay at the MAC layer (Marks, 2006).

We remark that the optional 802.16-2004 mesh mode is different from 802.16j. Actually, 802.16j is initiated to overcome mesh mode limitations because mesh mode replaces the PMP frame structure by point-to-point structure. Consequently, conventional 802.16-2204 PMP devices are not able to communicate with mesh devices. Thus, one of the main objectives of 802.16j is to design MMR without modifications to SSs. Hence, to retain the PMP backward compatible frame structure, 802.16j unlike mesh mode defines the network architecture to be tree based with BS as the root.

References

Alvrion Company, *Standards versus Proprietary Solutions: The Case for WiMAX Industry Standards*, April 25, 2005.

IEEE 802.16e Task Group, *IEEE Standard for Local and Metropolitan Area Networks—Part 16: Air Interface for Fixed and Mobile Broadband Wireless Access Systems—Amendment*

2: Physical and Medium Access Control Layers for Combined Fixed and Mobile Operation in Licensed Bands and Corrigendum 1, IEEE P802.16e/D12, February 2006.

IEEE License-Exempt (LE) Task Group, *Part 16: Air Interface for Fixed Broadband Wireless Access Systems—Amendment for Improved Coexistence Mechanisms for License-Exempt Operation*, IEEE Draft 802.16h, May 2006.

IEEE NetMan Task Group, *P802.16g Baseline Document to IEEE Standard for Local and Metropolitan Area Networks—Part 16: Air Interface for Fixed and Mobile Broadband Wireless Access Systems—Amendment to IEEE Standard for Local and Metropolitan Area Networks—Management Plane Procedures and Services*, August 2005a.

IEEE NetMan Task Group, *IEEE Standard for Local and Metropolitan Area Networks—Part 16: Air Interface for Fixed Broadband Wireless Access Systems—Amendment 1: Management Information Base*. IEEE Standard 802.16f, September 2005b.

IEEE NetMan Task Group, *P802.16k Draft Amendment to IEEE for Local and Metropolitan Area Networks: Media Access Control (MAC) Bridges—Amendment 2: Bridging of IEEE 802.16*, February 2006a.

IEEE NetMan Task Group, *Draft Amendment to IEEE Standard for Local and Metropolitan Area Networks—Part 16: Management Information Base Extensions*, P802.16i Baseline Document, October 2006b.

IEEE Relay Task Group, *P802.16j Amendment to IEEE Standard for Local and Metropolitan Area Networks—Part 16: Air Interface for Fixed and Mobile Broadband Wireless Access Systems—Physical and Medium Access Control Layers for Mobile Multihop Relay Specification*, March 2006.

David Johnston, *Bridging Support for 802.16*, IEEE 802.16k Presentation, Document number IEEE S802.16-06/001, March 2006.

LAN MAN Standards Committee of the IEEE Computer Society and the IEEE Microwave Theory and Techniques Society, *Local and Metropolitan Area Networks—Part 16: Air Interface for Fixed Broadband Wireless Access Systems*. Draft revision of IEEE Standard 802.16-2001. IEEE Standard 802.16-2001, 2002.

LAN MAN Standards Committee of the IEEE Computer Society and the IEEE Microwave Theory and Techniques Society, *Local and Metropolitan Area Networks—Part 16: Air Interface for Fixed Broadband Wireless Access Systems—amendment 1: Detailed System Profiles for 10–66 GHz*. IEEE Standard 802.16c-2002, 2002.

LAN MAN Standards Committee of the IEEE Computer Society and the IEEE Microwave Theory and Techniques Society, *Local and Metropolitan Area Networks—Part 16: Air Interface for Fixed Broadband Wireless Access Systems—amendment 2: Medium access control modifications and Additional Physical Layer specifications for 2–11 GHz*. IEEE Standard 802.16a-2003, 2003.

LAN MAN Standards Committee of the IEEE Computer Society and the IEEE Microwave Theory and Techniques Society, *IEEE Standard for Local and Metropolitan Area Networks—Part 16: Air Interface for Fixed Broadband Wireless Access Systems*, IEEE STD 802.16-2004, October 2004.

Roger B. Marks, *IEEE 802 Tutorial: 802.16 Mobile Multihop Relay*, March 2006.

MESA Project, *Technologies with Potential Applicability to Project MESA*, 2005.

Motorola Company, *WiMAX: E vs. D. The advantages of 802.16e over 802.16d*, 2005.

IEEE 802.16a Standard and WiMAX Igniting Broadband Wireless Access, 2004.

The WiMAX Forum, *Mobile WiMAX, Part I: A Technical Overview and Performance Evaluation*, 2006.

3

MAC Layer Protocol in WiMAX Systems

Maode Ma and Yan Zhang

CONTENTS

The medium access control (MAC) layer protocol of any communication system will normally describe or specify the issues of message composition and transmission, services provision and schemes, resources allocation, QoS support, and connection maintenance. This chapter will generally introduce the above issues at the MAC layer in WiMAX networks [1,2]. There are two types of topologies of the WiMAX system. One is the topology of point to multiple points (PMP) and the other is the mesh topology. We first introduce the operations and features of the two topologies and then describe the above issues in general.

3.1 Introduction

A network that utilizes a shared medium shall provide an efficient sharing mechanism. The PMP and mesh topology wireless networks are examples for sharing wireless media. The medium is radio waves in the space.

In the PMP mode of operation, the downlink, from the base station (BS) to subscriber stations (SSs), operates on a PMP basis. Within a given frequency channel and coverage of the BS sector, all SSs receive the same transmission or parts of it. The BS is the only transmitter operating in this direction. So it transmits without having to coordinate with other stations. The downlink is used for broadcasting the information. In cases where the message down link map (DL-MAP) does not explicitly indicate that a portion of the downlink subframe is for a specific SS, all SSs are able to listen to that portion. The SSs check the connection identifiers (CIDs) in the received protocol data units (PDUs) and retain only those PDUs addressed to them. SSs share the uplink to the BS on a demand basis. Depending on the class of service at the SSs, the SSs may be issued continuing rights to transmit or the transmission rights granted by the BS after receipt of requests from SSs. In addition to individually addressed messages, messages may also be sent by multicast to a group of selected SSs and broadcast to all SSs. In each sector, SSs are controlled by the transmission protocol at the MAC layer. And they are enabled to receive services to be tailored to the delay and bandwidth requirements of each application. It is accomplished by four types of uplink sharing schemes, which are unsolicited bandwidth grants, polling, and bandwidth requests contention.

The transmission scheme at the MAC layer is connection-oriented. All data communications are defined in the context of a connection. Service flows can be provisioned at an SS and connections are associated with these service flows, each of which is to provide transmission service at the requested bandwidth to a connection. The service flow defines the QoS parameters for the PDUs that are exchanged on the connection. The concept of a service flow on a connection is a key issue to the operation of the MAC protocol. Service flows provide a mechanism for uplink and downlink QoS management as bandwidth allocation processes. An SS requests uplink bandwidth on a per connection basis. Bandwidth is granted by the BS to an SS as an aggregate of grants in response to per connection requests from the SS. Connections may require active maintenance. And three connection management functions are supported by using static configuration and dynamic addition, modification, and deletion of connections. The termination of a connection is stimulated by the BS or SS.

Different from the PMP topology, in the operation of the mesh topology, traffic can occur directly between SSs and be routed through other SSs. The transmission can be managed by distributed scheduling, centralized scheduling, or a combination of both. Within a mesh network, a station that has a direct connection to backhaul services outside the mesh network is named a mesh BS. All the other stations of a mesh network are termed mesh SSs. Within mesh context, uplink and downlink are defined as traffic in the direction of the mesh BS and traffic away from the mesh BS, respectively. In a mesh network, there are neighbor, neighborhood, and extended neighborhood. The stations with direct links to a node are called neighbors of the node and neighbors of a node form a neighborhood. A node's neighbors are only one hop away from the node. An extended neighborhood contains all the neighbors of the neighborhood.

In a mesh system, every node including the mesh BS cannot transmit without having to coordinate with other nodes. By distributed scheduling, all the nodes shall coordinate their transmissions in their two-hop neighborhood and shall broadcast their schedules to all their neighbors. Optionally, the schedule may also be established by directed uncoordinated requests and grants between two nodes. Nodes shall ensure that the resulting transmissions do not cause collisions with the data and control traffic scheduled by any other node in the two-hop neighborhood. There is no difference in the mechanism for determining the schedule for downlink and uplink. By centralized scheduling, resources are granted in a more centralized manner. The mesh BS shall gather resource requests from all the mesh SSs within a certain hops range. It shall determine the amount of granted resources for each link in the network, both in downlink and uplink, and communicates these grants to all the mesh SSs within the hops range. Grant messages will not make any schedule, which should be determined by each node using a predetermined algorithm with given parameters.

All the communications are in the context of a link, which is established between two nodes. One link is used for all the data transmissions between the two nodes. QoS is provisioned over links on a message basis. No service or QoS parameters are associated with a link, but each unicast message has service parameters in the header. Traffic classification and flow regulation are performed at the ingress node by upper-layer classification/regulation protocol.

3.2 MAC Functions for the PMP Topology

This section will introduce the major parts of the MAC protocol specified in the IEEE802.16d standard, especially, for the functions and features of the MAC protocol to support PMP topology.

Inside a sector of the WiMAX systems each SS has a 48-bit universal MAC address, which uniquely defines the SS from within the set of all possible equipment types. It is used during the initial ranging process to establish the appropriate connections for an SS. It is also used as part of the authentication process for the BS and SS to verify each other. Connections are identified by a 16-bit CID. The CID is a connection identifier of the traffic at SSs, including connectionless traffic like IP, because it serves as a pointer to the destination and context information. Requests for transmission are based on these CIDs because the granted bandwidth may differ for different connections.

3.2.1   MAC PDU Composition and Transmission

3.2.1.1   MAC PDU Composition

Each MAC PDU is the basic unit of information prepared at the MAC layer and delivered to the physical layer. The PDU begins with a fixed-length generic MAC header. The header may be followed by the payload, which consists of zero or more subheaders and zero or more MAC service data units (SDUs) or fragments. The payload may vary in length so that a MAC PDU may represent a variable number of bytes. This allows the MAC to tunnel various higher-layer traffic types without any knowledge of the formats or bit patterns of those messages.

There are two types of MAC headers. The first type is the generic MAC header in each MAC PDU containing either MAC management messages or data made at the convergence layer. The second type is the bandwidth request header for requesting additional bandwidth. Five types of subheaders may be inserted in MAC PDUs immediately following the Generic MAC header. The mesh subheader could exist before all the other subheaders. After this, the Grant Management subheader will come next. And the FAST FEEDBACK Allocation subheader always appears as the last per-PDU subheader. The Packing and Fragmentation subheaders are mutually exclusive and both will not be present in the same MAC PDU. A set of MAC management messages are defined. These messages are carried in the Payload of the MAC PDU. All MAC Management messages begin with a Management Message Type field and may contain additional fields.

Multiple MAC PDUs could be concatenated into a single transmission unit in either the uplink or downlink. Since each MAC PDU is identified by a unique CID, the receiving MAC entity is able to present the MAC SDU (after reassembling the MAC SDU from one or more received MAC PDUs) to the correct instance of the MAC service access point (SAP). MAC Management messages, user data, and bandwidth request MAC PDUs may be concatenated into the same transmission.

Fragmentation is the process by which a MAC SDU is divided into one or more MAC PDUs. This process is undertaken to allow efficient use of available bandwidth relative to the QoS requirements of a connection's service flow. Capabilities of fragmentation and reassembly are mandatory. The authority to fragment traffic on a connection is defined when the connection is created by the MAC SAP. Fragmentation may be initiated by a BS for downlink connections and by an SS for uplink connections.

The MAC protocol can pack multiple MAC SDUs into a single MAC PDU. Packing makes use of the connection attribute indicating whether the connection carries fixed-length or variable-length packets. For packing with fixed-length blocks, the request/transmission policy shall be set to allow packing and prohibit fragmentation, and the SDU size shall be included in dynamic service activate request (DSA-REQ) message when establishing the connection. The length field of the MAC header implicitly indicates the number of MAC SDUs packed into a single MAC PDU. When packing variable-length SDU connections, the indication of where one MAC SDU ends and another begins is necessary. In the variable-length MAC SDU case, the MAC attaches a Packing subheader to each MAC SDU.

3.2.1.2 MPDU Transmission

At the MAC layer, MAC protocol data unit (MPDU) transmission is supported. The following issues support the MPDU transmission.

3.2.1.2.1 Duplex Techniques

Several duplexing techniques are supported by the MAC protocol. The choice of duplexing technique may affect certain physical layer (PHY) parameters as well as impact the features that can be supported.

In an frequency division duplex (FDD) system, the uplink and downlink channels are located on separate frequencies and the downlink data can be transmitted in bursts. A fixed duration frame is used for both uplink and downlink transmissions. This facilitates the use of different modulation types. It also allows simultaneous use of both full-duplex SSs and, optionally, half-duplex SSs. If half-duplex SSs are used, the bandwidth controller shall not allocate uplink bandwidth for a half-duplex SS at the same time that it is expected to receive data on the downlink channel, including allowance for the propagation delay, SS transmit/receive transition gap (SSTTG), and SS receive/transmit transition gap (SSRTG). The fact that the uplink and downlink channels utilize a fixed duration frame simplifies the bandwidth allocation algorithms. A full-duplex SS is capable of continuously listening to the downlink channel, while a half-duplex SS can listen to the downlink channel only when it is not transmitting in the uplink channel.

In the case of time division duplex (TDD), the uplink and downlink transmissions occur at different times and usually share the same frequency. A TDD frame has a fixed duration and contains one downlink and one uplink subframe. The frame is divided into an integer number of physical slots (PSs), which help to partition the bandwidth easily. The TDD framing is adaptive in that the bandwidth allocated to the downlink versus the uplink can vary. The split between uplink and downlink is a system parameter and is controlled at higher layers within the system.

The DL-MAP message defines the usage of the downlink intervals for a burst mode PHY. The uplink bandwidth allocation map (UL-MAP) defines the uplink usage in terms of the offset of the burst relative to the allocation start time.

3.2.1.2.2   Uplink Timing and Allocations

Uplink timing is referenced from the beginning of the downlink subframe. The allocation start time in the UL-MAP is referenced from the start of the downlink subframe and may be such that the UL-MAP references some point in the current or a future frame. The SS shall always adjust its concept of uplink timing based upon the timing adjustments sent in the ranging response (RNG-RSP) messages.

For the single carrier (SC) and single carrier access (SCa) PHY layers, the UL-MAP uses units of minislots. The size of the minislot is specified as a function of PSs and is carried in the upper link channel descriptor (UCD) for each uplink channel. For the orthogonal frequency division multiplexing (OFDM) and orthogonal frequency division multiple access (OFDMA) PHY layers, the UL-MAP uses units of symbols and subchannels.

Through the request IE, the BS specifies an uplink interval in which requests may be made for bandwidth and for uplink data transmission. The character of this IE changes depending on the type of CID used in the IE. If broadcast or multicast, this is an invitation for SSs to contend for requests. If unicast, this is an invitation for a particular SS to request bandwidth. Unicasts may be used as part of a QoS scheduling scheme. For any uplink allocation, the SS may optionally decide to use the allocation for data or requests (or requests piggybacked in data). PDUs transmitted in this interval shall use the bandwidth request header format. For bandwidth request contention opportunities, the BS shall allocate a grant that is an integer multiple of the value of "Bandwidth request opportunity size," which shall be published in each UCD transmission.

Timing information in the DL-MAP and UL-MAP is relative. The following time instants are used as a reference for timing information: (1) DL-MAP: The start of the first symbol (including the preamble if present) of the frame in which the message was transmitted. (2) UL-MAP: The start of the first symbol (including the preamble if present) of the frame in which the message was transmitted plus the value of the allocation start time. Information in the DL-MAP pertains to the current frame (the frame in which the message was received). Information carried in the UL-MAP pertains to a time interval starting at the allocation start time measured from the beginning of the current frame and ending after the last specified allocation. This timing holds for both the TDD and FDD variants of operation.

3.2.1.3   MPDU Retransmission Scheme

The automatic retransmission (ARQ) mechanism is a part of the MAC, which is optional for implementation. When implemented, ARQ may be enabled on a per connection basis. The ARQ shall be specified and negotiated during

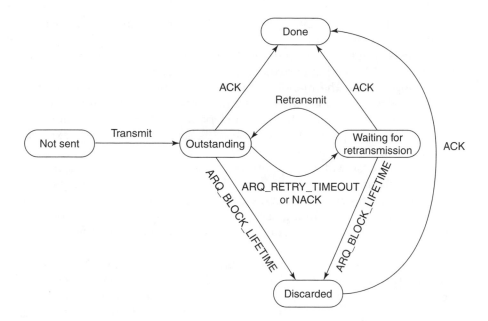

FIGURE 3.1
Operations of the ARQ scheme.

connection creation. A connection cannot have a mixture of ARQ and non-ARQ traffic. Similar to other properties of the MAC protocol, the scope of a specific instance of ARQ is limited to one unidirectional connection.

For ARQ-enabled connections, enabling of fragmentation is optional. When fragmentation is enabled, the transmitter may partition each MAC SDU into fragments for separate transmission based on the value of the ARQ_BLOCK_SIZE parameter. When fragmentation is not enabled, the connection shall be managed as if fragmentation was enabled. In this case, regardless of the negotiated block size, each fragment formed for transmission shall contain all the blocks of data associated with the parent MAC SDU. The ARQ feedback information can be sent as a stand-alone MAC management message on the appropriate basic management connection, or piggybacked on an existing connection. ARQ feedback cannot be fragmented (Figure 3.1).

3.2.2 Services Provision and Schemes

3.2.2.1 Services and Parameters

Scheduling services represent the data handling mechanisms supported by the MAC scheduler for data transport on a connection. Each connection is associated with a single data service. Each data service is associated with a set of QoS parameters that quantify aspects of its behavior. These parameters are managed using the dynamic service addition (DSA) and dynamic service change (DSC) message dialogs. Four services are supported: unsolicited grant service (UGS), real-time polling service (rtPS), nonreal-time polling service (nrtPS), and best effort (BE).

The UGS is designed to support real-time data streams consisting of fixed-size data packets issued at periodic intervals, such as voice over IP without silence suppression. The mandatory QoS service flow parameters for this scheduling service are maximum sustained traffic rate, maximum latency, tolerated jitter, and request/transmission policy. If present, the minimum reserved traffic rate parameter shall have the same value as the maximum sustained traffic rate parameter.

The rtPS is designed to support real-time data streams consisting of variable-sized data packets that are issued at periodic intervals, such as moving pictures experts group (MPEG) video. The mandatory QoS service flow parameters for this scheduling service are minimum reserved traffic rate, maximum sustained traffic rate, maximum latency, and request/transmission policy.

The nrtPS is designed to support delay-tolerant data streams consisting of variable-sized data packets for which a minimum data rate is required, such as FTP. The mandatory QoS service flow parameters for this scheduling service are minimum reserved traffic rate, maximum sustained traffic rate, traffic priority, and request/transmission policy.

The BE service is designed to support data streams for which no minimum service level is required and therefore may be handled on a space-available basis. The mandatory QoS service flow parameters for this scheduling service are maximum sustained traffic rate, traffic priority, and request/transmission policy.

3.2.2.2 *Service Implementation Schemes*

3.2.2.2.1 *Uplink Scheduling Scheme*

Uplink request/grant scheduling is performed by the BS with the intention to provide each SS with bandwidth for uplink transmissions or opportunities to request bandwidth. By specifying a scheduling service and its associated QoS parameters, the BS scheduler can anticipate the throughput and latency needs of the uplink traffic and provide polls or grants at the appropriate times.

3.2.2.2.1.1 UGS Service

The UGS is designed to support real-time service flows that generate fixed-size data packets on a periodic basis, such as T1/E1 and voice over IP without silence suppression. The service offers fixed-size grants on a real-time periodic basis, which eliminate the overhead and latency of SS requests and assure that grants are available to meet the flow's real-time needs. The BS shall provide data grant burst IEs to the SS at periodic intervals based upon the maximum sustained traffic rate of the service flow. The size of these grants shall be sufficient to hold the fixed-length data associated with the service flow but may be larger at the discretion of the BS scheduler.

For this service to work correctly, the request/transmission policy setting shall be such that the SS is prohibited from using any contention request opportunities for this connection. The key service IEs are the maximum sustained traffic, maximum latency, the tolerated jitter, and the request/transmission policy. If present, the minimum reserved traffic rate parameter shall have the same value as the maximum sustained traffic rate parameter.

The grant management subheader is used to pass status information from the SS to the BS regarding the state of the UGS service flow. The most significant bit of the grant management field is the slip indicator (SI) bit. The SS shall set this flag once it detects that this service flow has exceeded its transmit queue depth. Once the SS detects that the service flow's transmission queue is back within limits, it shall clear the SI flag. The flag allows the BS to provide for long-term compensation for conditions such as lost maps or clock rate mismatches by issuing additional grants. The poll-me (PM) bit may be used to request to be polled for a different, non-UGS connection.

The BS shall not allocate more bandwidth than the maximum sustained traffic rate parameter of the active QoS parameter set, excluding the case when the SI bit of the grant management field is set. In this case, the BS may grant up to 1% additional bandwidth for clock rate mismatch compensation.

3.2.2.2.1.2 rtPS Service

The rtPS is designed to support real-time service flows that generate variable-size data packets on a periodic basis, such as MPEG video. The service offers real-time, periodic, unicast request opportunities, which meet the flow's real-time needs and allow the SS to specify the size of the desired grant. This service requires more request overhead than UGS, but supports variable grant sizes for optimum data transport efficiency.

The BS shall provide periodic unicast request opportunities. For this service to work correctly, the request/transmission policy setting shall be such that the SS is prohibited from using any contention request opportunities for that connection. The BS may issue unicast request opportunities as prescribed by this service even if prior requests are currently unfulfilled. This results in the SS using only unicast request opportunities to obtain uplink transmission opportunities (the SS could still use unsolicited data grant burst types for uplink transmission as well). All other bits of the request/transmission policy are irrelevant to the fundamental operation of this scheduling service and should be set according to network policy. The key service IEs are the maximum sustained traffic rate, the minimum reserved traffic rate, the maximum latency, and the request/transmission policy.

3.2.2.2.1.3 nrtPS Service

The nrtPS offers unicast polls on a regular basis, which assures that the service flow receives request opportunities even during network congestion. The BS typically polls nrtPS CIDs on an interval on the order of one second or less.

The BS shall provide timely unicast request opportunities. For this service to work correctly, the request/transmission policy setting shall be set such that the SS is allowed to use contention request opportunities. This results in the SS using contention request opportunities as well as unicast request opportunities and unsolicited data grant burst types. All other bits of the request/transmission policy are irrelevant to the fundamental operation of this scheduling service and should be set according to network policy.

3.2.2.2.1.4   BE Service

The intent of the BE service is to provide efficient service for best effort traffic. For this service to work correctly, the request/transmission policy setting shall be set such that the SS is allowed to use contention request opportunities. This results in the SS using contention request opportunities as well as unicast request opportunities and unsolicited data grant burst types. All other bits of the request/transmission policy are irrelevant to the fundamental operation of this scheduling service and should be set according to network policy.

3.2.2.2.2   *Bandwidth Allocation Scheme*

During network entry and initialization, every SS is assigned up to three dedicated CIDs for the purpose of sending and receiving control messages. These connection pairs are used to allow differentiated levels of QoS service to be applied to the different connections carrying MAC management traffic. Changing bandwidth requirements is necessary for all services except constant bit rate UGS connections. Demand assigned multiple access (DAMA) services will provide resources on a demand assignment basis, as the need arises. When an SS needs to ask for bandwidth on a connection with BE scheduling service, it sends a message to the BS containing the immediate requirements of the DAMA connection. QoS for the connection was established at connection establishment and is looked up by the BS. There are numerous methods by which the SS can get the bandwidth request message to the BS.

3.2.2.2.2.1   Requests

Requests are for SSs to indicate to the BS that they need uplink bandwidth allocation. A request may come as a stand-alone bandwidth request header or it may come as a piggyback request. As the uplink burst profile can change dynamically, all requests for bandwidth shall be made in terms of the number of bytes needed to carry the MAC header and payload, but not the PHY overhead. The bandwidth request message may be transmitted during an uplink allocation except during an initial ranging interval. Bandwidth requests may be incremental or aggregate. When the BS receives an incremental bandwidth request, it shall add the quantity of bandwidth requested to its current perception of the bandwidth needs of the connection. When the BS receives an aggregate bandwidth request, it shall replace its perception of the bandwidth needs of the connection with the quantity of bandwidth requested. The piggybacked bandwidth requests shall always be incremental. The self-correcting nature of the request/grant protocol requires that the SSs shall periodically use aggregate bandwidth requests. The period may be a function of the QoS of a service and of the link quality. Owing to the possibility of collisions, bandwidth requests transmitted in broadcast or multicast request IEs should be aggregate requests.

3.2.2.2.2 Grants

For an SS, bandwidth requests are not to individual connections while each bandwidth grant is addressed to the SS's basic CID. In all cases, based on the latest information received from the BS and the status of the request, the SS may decide to perform backoff and request again or to discard the MAC SDU. An SS may use request IEs that are broadcast, directed at a multicast polling group it is a member of, or directed at its basic CID. In all cases, the request IE burst profile is used, even if the BS is capable of receiving the SS with a more efficient burst profile. To take advantage of a more efficient burst profile, the SS should transmit in an interval defined by a data grant IE directed at its basic CID. Owing to this, unicast polling of an SS would normally be done by allocating a data grant IE directed at its basic CID. Also note that in a data grant IE directed at its basic CID, the SS may make bandwidth requests for any of its connections.

3.2.2.2.3 *Request Transmission Schemes*

There are two ways to issue the bandwidth requests. In the rtPS and nrtPS services, the requests will be issued by the control of polling scheme or contention. In the BE service, the requests will be issued mainly by contention.

3.2.2.2.3.1 Polling

Polling is the process by which the BS allocates to the SSs bandwidth specifically for the purpose of making bandwidth requests. These allocations may be to individual SSs or to groups of SSs. Allocations to groups of connections or SSs actually define bandwidth request contention IEs. The allocations are not in the form of an explicit message but are contained as a series of IEs within the UL-MAP. Polling is done on SS basis. Bandwidth is always requested on a connection basis and bandwidth is allocated on an SS basis.

When an SS is polled individually, it is the unicast polling scheme without an explicit message that is transmitted to poll the SS. Rather, the SS is allocated, in the UL-MAP, bandwidth sufficient to respond with a bandwidth request. If the SS does not need bandwidth, the allocation is padded. SSs that have an active UGS connection of sufficient bandwidth shall not be polled individually unless they set the PM bit in the header of a packet on the UGS connection. This saves bandwidth over polling all SSs individually. Note that unicast polling would normally be done on a per-SS basis by allocating a data grant IE directed at its basic CID.

If insufficient bandwidth is available to individually poll many inactive SSs, some SSs may be polled in multicast groups or a broadcast poll may be issued. As with individual polling, the poll is not an explicit message, but bandwidth allocated in the UL-MAP. The difference is that, rather than associating allocated bandwidth with an SS's basic CID, the allocation is to a multicast or broadcast CID.

When the poll is directed at a multicast or broadcast CID, an SS belonging to the polled group may request bandwidth during any request interval allocated to that CID in the UL-MAP by a request IE. To reduce the likelihood of collision with multicast and broadcast polling, only SS's needing bandwidth reply. They shall take the contention resolution algorithm to select the time slot in which to transmit the initial bandwidth request. The SS shall assume that the transmission has been unsuccessful if no grant has been received in the number of subsequent UL-MAP messages specified by the parameter contention-based reservation timeout. Note that, with a frame-based PHY with UL-MAPs occurring at predetermined instants, erroneous UL-MAPs may be counted towards this number. If the request is made in a multicast or broadcast opportunity, the SS continues to run the contention resolution algorithm.

3.2.2.2.3.2   Contention Resolution

The mandatory contention resolution method is the truncated binary exponential backoff with the initial backoff window and the maximum backoff window controlled by the BS. When an SS has information to send and wants to enter the contention resolution process, it sets its internal backoff window equal to the request backoff start defined in the UCD message referenced by the UCD count in the UL-MAP message currently in effect. The SS shall randomly select a number within its backoff window. This random value indicates the number of contention transmission opportunities that the SS shall defer before transmitting. An SS shall consider only contention transmission opportunities for which this transmission would have been eligible. These are defined by request IEs in the UL-MAP messages.

The SS shall now increase its backoff window by a factor of two, as long as it is less than the maximum backoff window. The SS shall randomly select a number within its new backoff window and repeat the deferring process described above. This retry process continues until the maximum number (i.e., request retries for bandwidth requests and contention ranging retries for initial ranging) of retries has been reached. At this time, for bandwidth requests, the PDU shall be discarded.

For bandwidth requests, if the SS receives a unicast request IE or data grant burst type IE at any time while deferring for this CID, it shall stop the contention resolution process and use the explicit transmission opportunity. The BS has much flexibility in controlling the contention resolution. At one extreme, the BS may choose to set up the request (or ranging) backoff start and request (or ranging) backoff end to emulate an Ethernet-style backoff with its associated simplicity and distributed nature as well as its fairness and efficiency issues.

A transmission opportunity is defined as an allocation provided in a UL-MAP or part thereof intended for a group of SSs authorized to transmit bandwidth requests or initial ranging requests. This group may include either all SSs having an intention to join the cell or all registered SSs or a multicast polling group. The number of transmission opportunities associated with a particular IE in a map is dependent on the total size of the allocation as well as the size of an individual transmission. The size of an individual transmission opportunity for each type of contention IE shall be published in each transmitted UCD message. The BS shall always allocate bandwidth for contention IEs in integer multiples of these published values (Figure 3.2).

3.2.3 Connection Establishment and Maintenance

3.2.3.1 Network Entry and Initialization

This is the first step when a new SS enters and registers to one sector of the WiMAX network with the PMP operation.

The procedure can be divided into the following phases:

(a) Scan for downlink channel and establish synchronization with the BS
(b) Obtain transmit parameters (from UCD message)
(c) Perform ranging
(d) Negotiate basic capabilities
(e) Authorize SS and perform key exchange
(f) Perform registration
(g) Establish IP connectivity
(h) Establish time of day
(i) Transfer operational parameters
(j) Set up connections

Implementation of phases (g), (h), and (i) at the SS is optional. These phases shall only be performed if the SS has indicated in the registration request (REG-REQ) message that it is a managed SS. Each SS contains the following information when shipped from the manufacturer: (a) A 48-bit universal MAC address assigned during the manufacturing process. This is used to identify the SS to the various provisioning servers during initialization. (b) Security information used to authenticate the SS to the security server and authenticate the responses from the security and provisioning servers.

3.2.3.2 Connection Maintenance

Ranging is a collection of processes by which the SS and BS maintain the quality of the RF communication link between them. Distinct processes are used for managing uplink and downlink. Also, some PHY modes support ranging mechanisms unique to their capabilities.

The channel descriptors are transmitted at regular intervals by the BS. Each descriptor contains the configuration change count, which shall remain

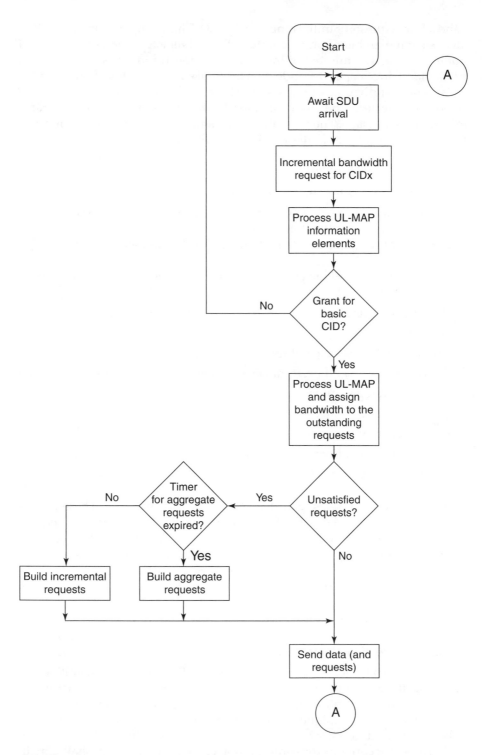

FIGURE 3.2
SS request/grant procedure.

unchanged as long as the channel descriptor remains unchanged. All UL-MAP and DL-MAP messages allocating transmissions and receptions using burst profiles defined in a channel descriptor with a given configuration change count value shall have a UCD/DCD count value equal to the configuration change count of the corresponding channel descriptor. The BS may add an SS to a multicast polling group by sending an MCA-REQ message with the join command. Upon receiving an MCA-REQ message, the SS shall respond by sending an MCA-RSP message.

The BS may establish a downlink multicast service by creating a connection with each SS to be associated with the service. Any available traffic CID value may be used for the service. To ensure proper multicast operation, the CID used for the service is the same for all SSs on the same channel that participate in the connection. The SSs need not be aware that the connection is a multicast connection. The data transmitted on the connection with the given CID shall be received and processed by the MAC of each involved SS. Thus, each multicast SDU is transmitted only once per BS channel. Since a multicast connection is associated with a service flow, it is associated with the QoS and traffic parameters for that service flow.

ARQ is not applicable to multicast connections. If a downlink multicast connection is to be encrypted, each SS participating in the connection shall have an additional security association (SA), allowing that connection to be encrypted using keys that are independent of those used for other encrypted transmissions between the SSs and the BS.

3.2.4 QoS Services

There are several QoS related concepts defined in the IEEE 802.16 standards. These concepts cover the following: service flow QoS scheduling, dynamic service establishment, and two-phase activation model.

The principal mechanism for providing QoS is to associate packets traversing the MAC interface into a service flow as identified by the transport CID. A service flow is a unidirectional flow of packets that is provided a particular QoS. The SS and BS provide this QoS according to the QoS parameter set defined for the service flow. Service flows exist in both the uplink and downlink direction and may exist without actually being activated to carry traffic. All service flows have a 32-bit service flow identified (SFID); admitted and active service flows also have a 16-bit CID.

The primary purpose of the QoS features is to define transmission ordering and scheduling on the air interface. However, these features often need to work in conjunction with mechanisms beyond the air interface to provide end-to-end QoS or to police the behavior of SSs. So, the key requirements for QoS are listed as follows:

(a) A configuration and registration function for preconfiguring SS-based QoS service flows and traffic parameters.

(b) A signaling function for dynamically establishing QoS-enabled service flows and traffic parameters.

(c) Utilization of MAC scheduling and QoS traffic parameters for uplink service flows.

(d) Utilization of QoS traffic parameters for downlink service flows.

(e) Grouping of service flow properties into named service classes, so upper-layer entities and external applications (at both the SS and BS) may request service flows with the desired QoS parameters in a globally consistent way.

A service flow is a MAC transport service that provides unidirectional transport of packets either to uplink packets transmitted by the SS or to downlink packets transmitted by the BS. A service flow is characterized by a set of QoS parameters such as latency, jitter, and throughput assurances. To standardize operation between the SS and BS, these attributes include details of how the SS requests uplink bandwidth allocations and the expected behavior of the BS uplink scheduler.

To most efficiently utilize network resources such as bandwidth and memory, 802.16 adopts a two-phase activation model in which resources assigned to a particular admitted service flow may not be actually committed until the service flow is activated. Each admitted or active service flow is mapped to a MAC connection with a unique CID. Generally, there are three basic types of service flows, namely

(a) Provisioned service flows: This service flow may be provisioned but not immediately activated and defers admission. The network assigns a SFID for such a service flow. The BS may also require an exchange with a policy module prior to admission.

(b) Admitted service flows: This protocol supports a two-phase activation model that is often utilized in telephony applications. In the two-phase activation model, the resources are first "admitted" and once the end-to-end negotiation is completed, the resources are "activated." The two-phase model helps to conserve network resources until a complete end-to-end connection has been established. It performs policy checks and admission control on resources as quickly as possible and, in particular, before informing the far end of a connection request, preventing several potential theft-of-service scenarios.

(c) Active service flows: A service flow that has a non-NULL ActiveQoSParamSet is said to be an active service flow. It is requesting according to its request/transmission policy and being granted bandwidth for transport of data packets. An admitted service flow may be activated by providing an ActiveQoSParamSet, signaling the resources actually desired at the current time. This completes the second stage of the two-phase activation model.

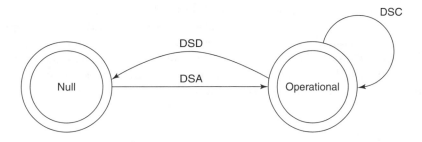

FIGURE 3.3
Overview of dynamic service flow.

IEEE 802.16 also supports dynamic service changes in which service flow parameters are renegotiated. Like dynamic service flow establishment, service flow changes also follow a similar three-way handshaking protocol. Service flows may be created, changed, or deleted. This is accomplished through a series of MAC management messages referred to as DSA, DSC, and dynamic service delete (DSD). The DSA messages create a new service flow, the DSC messages change an existing service flow, and the DSD messages delete an existing service flow (Figure 3.3).

In general, service flows in IEEE 802.16 are preprovisioned and setup of the service flows is initiated by the BS during SS initialization. However, service flows can also be dynamically established and immediately activated by either the BS or the SS. The SS typically initiates service flows only if there is a dynamically signaled connection, such as a switched virtual connection (SVC) from an ATM network. The establishment of service flows is performed through a three-way handshaking protocol in which the request for service flow establishment is responded to and the response acknowledged.

3.3 MAC Functions for the Mesh Topology

In this section, we will focus on features and functions provided by the MAC layer protocol to support WiMAX mesh networks. Although the mesh topology has its distinct characteristics, some basic functions provided by the MAC protocol for the PMP topology are applicable in the mesh topology. This section will provide a general overview on the MAC protocol support to the mesh topology.

3.3.1 Addressing and Connections

For addressing nodes in the local neighborhood, 8-bit link identifiers (link IDs) shall be used. Each node shall assign an ID for each link it has established with its neighbors. The link IDs are communicated during the link establishment process as neighboring nodes establish new links. The link ID is transmitted as part of the CID in the generic MAC header in unicast messages. The link

IDs shall be used in distributed scheduling to identify resource requests and grants. Since these messages are broadcast, the receiver nodes can determine the schedule using the transmitter's node ID in the mesh subheader, and the link ID in the payload of the mesh mode schedule with distributed scheduling (MSH-DSCH) message. The connection ID in mesh mode is specified to convey broadcast/unicast, service parameters, and the link identification.

3.3.2 Bandwidth Allocation

Unlike the PMP mode, there are no clearly separate downlink and uplink subframes in the mesh mode. Each station is able to create direct communication links with a number of other stations in the network instead of communicating only with a BS. However, in typical installations there will still be certain nodes that provide the BS function of connecting the mesh network to the backhaul links. In fact, when using mesh-centralized scheduling, these BS nodes perform much of the same basic functions as do the BS in PMP mode. Thus, the key difference is that in mesh mode all the SSs may have direct links with other SSs. Further, there is no need to have a direct link from an SS to the BS of the mesh network. This connection can be provided through other SSs. Communication in all these links shall be controlled by a centralized algorithm scheduled in a distributed manner within each node's extended neighborhood, or scheduled using a combination of these.

3.3.2.1 *Distributed Scheduling*

The stations that have direct links are called neighbors and shall form a neighborhood. A node's neighbors are considered to be "one hop" away from the node. A two-hop extended neighborhood contains, additionally, all the neighbors of the neighborhood. In the coordinated distributed scheduling mode, all the stations (BS and SSs) shall coordinate their transmissions in their extended two-hop neighborhood.

The coordinated distributed scheduling mode uses some or the entire control portion of each frame to regularly transmit its own schedule and proposed schedule changes on a PMP basis to all its neighbors. Within a given channel, all neighboring stations receive the same schedule transmissions. All the stations in a network shall use this same channel to transmit schedule information in a format of specific resource requests and grants. Coordinated distributed scheduling ensures that transmissions are scheduled in a manner that does not rely on the operation of a BS, and that are not necessarily directed to or from the BS.

Within the constraints of the coordinated schedules (distributed or centralized), uncoordinated distributed scheduling can be used for fast, *ad hoc* setup of schedules on a link-by-link basis. Uncoordinated distributed schedules are established by directed requests and grants between two nodes and shall be scheduled to ensure that the resulting data transmissions (and the request and grant packets themselves) do not cause collisions with the data and control

traffic scheduled by the coordinated distributed or the centralized scheduling methods. Both the coordinated and uncoordinated distributed scheduling
employ a three-way handshake.

- MSH-DSCH: Request is made along with MSH-DSCH: Availabilities, which indicate potential slots for replies and actual schedule.
- MSH-DSCH: Grant is sent in response indicating a subset of the
  suggested availabilities that fits, if possible, the request. The neighbors of this node not involved in this schedule shall assume that the
  transmission takes place as granted.
- MSH-DSCH: Grant is sent by the original requester containing a
  copy of the grant from the other party, to confirm the schedule to the
  other party. The neighbors of this node not involved in the schedule
  shall assume that the transmission takes place as granted.

The differences between coordinated and uncoordinated distributed scheduling are as follows: In the coordinated case, the MSH-DSCH messages are
scheduled in the control subframe in a collision-free manner; whereas, in the
uncoordinated case, MSH-DSCH messages may collide. Nodes responding
to a request should, in the uncoordinated case, wait a sufficient number of
minislots of the indicated availabilities before responding with a grant, such
that nodes listed earlier in the request have an opportunity to respond. The
grant confirmation is sent in the minislots immediately following the first
successful reception of an associated grant packet.

3.3.2.2 Centralized Scheduling

The schedule using centralized scheduling is determined in a centralized
manner than in the distributed scheduling mode. The network connections
and topology are the same as in the distributed scheduling mode. However,
the scheduled transmissions for the SSs shall be defined by the BS. The BS
determines the flow assignments from the resource requests from the SSs. Subsequently, the SSs determine the actual schedule from these flow assignments
by using a common algorithm that divides the frame proportionally to the
assignments. Thus, the BS acts just like the BS in a PMP network except that
not all of the SSs have to be directly connected to the BS, and the assignments
determined by the BS extends to those SSs not directly connected to the BS.
The SS resource requests and the BS assignments are both transmitted during
the control portion of the frame. Centralized scheduling ensures that transmissions are coordinated to ensure collision-free scheduling over the links in
the routing tree to and from the BS, typically in a more optimal manner than
the distributed scheduling method for traffic streams (or collections of traffic
streams that share links), which persist over a duration that is greater than
the cycle time to relay the new resource requests and distribute the updated
schedule.

Only TDD is supported in mesh mode. Contrary to the basic PMP mode, there are no clearly separate downlink and uplink subframes in the mesh mode. Stations shall transmit to each other either in scheduled channels or in random access channels as in PMP mode. All the basic functions like scheduling and network synchronization are based on the neighbor information that all the nodes in the mesh network shall maintain. Each node (BS or SS) maintains a physical neighborhood list.

When using coordinated distributed scheduling, all the stations in a network shall use the same channel to transmit schedule information in a format of specific resource requests and grants in MSH-DSCH messages. A station shall indicate its own schedule by transmitting a MSH-DSCH regularly. The MSH-DSCH messages shall be transmitted during the control portion of the frame. An SS that has a direct link to the BS shall synchronize to the BS while an SS that is at least two hops from the BS shall synchronize to its neighbor SSs that are closer to the BS.

When using centralized scheduling, the BS shall act as a centralized scheduler for the SSs. Using centralized scheduling, the BS shall provide schedule configuration (MSH-CSCF) and assignments (MSH-CSCH) to all SSs. The BS determines the assignments from the resource requests received from the SSs. Intermediate SSs are responsible for forwarding these requests for SSs (listed in the current routing tree as specified by the last MSH-CSCF modified by the last MSH-CSCH update) that are further from the BS (i.e., more hops from the BS) as needed. All the SSs shall listen and compute the schedule. Further, they shall forward the MSH-CSCH message to their neighbors that are further away from the BS.

3.3.2.3 Mesh Network Synchronization

Network configuration (MSH-NCFG) and network entry (MSH-NENT) packets provide a basic level of communication among nodes in different nearby networks, whether from the same or different equipment vendors or wireless operators. These packets are used to synchronize both centralized and distributed control mesh networks. This communication is used to support basic configuration activities such as synchronization between nearby networks used (i.e., for multiple, colocated BSs to synchronize their uplink and downlink transmission periods), communication and coordination of channel usage by nearby networks, and discovery and basic network entry of new nodes.

MSH-NCFG, MSH-NENT, and MSH-DSCH can assist a node in synchronizing to the start of frames. For these messages, the control subframe, which initiates each frame, is divided into transmit opportunities. The first transmit opportunity in a network control subframe may only contain MSH-NENT messages, while the remainder MSH-CTRL-LEN-1 may only contain MSH-NCFG messages. In scheduling control subframes, the MSH-DSCH-NUM transmit opportunities assigned for MSH-DSCH messages come last in the control subframe. The MSH-NCFG messages also contain the number of its transmit opportunity, which allows nodes to easily calculate the start time of the frame.

3.4 Summary

In this chapter, we have reviewed the functions and features of the core MAC protocol of the WiMAX systems including the PMP topology and mesh topology. In the standard, the MAC protocol should include another two sublayers, which are convergence sublayer and security sublayer. However, they have not been covered in this chapter. Only the fundamental part of the MAC protocol of the WiMAX systems has been summarized and presented. As a part of communication protocol stack, MAC protocol plays a very important role in the communication procedure. And this is the reason why MAC protocol has been specified in almost every communication standard by the IEEE standard committee. This chapter is expected to be a carrier of the fundamental knowledge of the MAC protocol specified in the IEEE 802.16d and the understanding of the features and functions of the MAC protocols for the WiMAX systems.

References

1. IEEE 802.16-2004, *IEEE Standard for Local and Metropolitan Area Networks—Part 16: Air Interface for Fixed Broadband Wireless Access Systems*, October 1, 2004.
2. Frank Ohrtman, *WiMax Handbook: Building 802.16 Wireless Networks*, McGraw-Hill, New York, USA, 2005.

4

Scheduling and Performance Analysis of QoS for IEEE 802.16 Broadband Wireless Access Network

James T. Yu

CONTENTS

4.1   Introduction

Over the last few years, we have seen the continual growth and demand for broadband wireless access (BWA) for residential, business, and mobile customers. With the standardization of IEEE 802.16-2004 [1], the industry formed the WiMAX Forum to support product certification of conformance to the standard and to promote interoperability among different vendors' products. A recent data shows 48% increase in the WiMAX equipment market, from $45M in fourth quarter of 2005 to $70M in first quarter of 2006 [2]. The same report predicts that the WiMAX market will grow to more than $1B by 2009. The success of WiFi (as specified in 802.11) is evident on the wireless local area network (WLAN), and many people expect to see the same growth of WiMAX on the wide area network (WAN). In general, LAN is based on broadcast technology with connectionless services, and WAN is based on point-to-point (P2P) or point-to-multipoint (P2MP) technology with connection-oriented services.

   Supporting Quality of Service (QoS) is essential for WAN because it allows for more efficient operations on the service providers' network to meet various customer demands. A service provider can offer differentiated services with specific Service Level Agreement (SLA) and charge the services accordingly. The IEEE 802.16 standard supports QoS on a per *connection* basis, where a connection is defined between the base station (BS) and a subscriber station (SS). A connection could be either from the BS to an SS (a downlink or DL connection) or from an SS to the BS (an uplink or UL connection). An SS could establish multiple connections to the BS, where each connection has its own QoS. An SS requests for bandwidth allocation on a DL or UL channel and the BS allocates the bandwidth to the SS based on the available resources, which is in the radio frequency spectrum. After granting the bandwidth, the BS enters the request into a priority queue based on its QoS. The BS then applies a scheduling algorithm to determine when and how to serve the jobs in the queues. The IEEE 802.16 standard provides a protocol for the request/grant procedure. However, the standard does not provide the QoS scheduling algorithm and its implementation is open to the product vendors. In this chapter, we present an architecture and an operation procedure of admission control and job scheduling, and develop a simulation model to study the network performance under various load conditions.

4.2   Background

4.2.1   QoS Definition

QoS is the guarantee of the service-level performance for a data stream from a source to a destination [3]. Such an assurance, of course, shall not exceed the

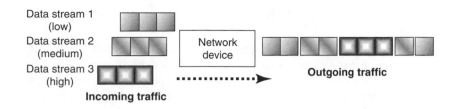

FIGURE 4.1
Need for QoS.

physical capacity of the transmission media. For example, we cannot guarantee a throughput of 100 Mbps on a Cat-3 cable that supports only 10 Mbps. Also, we cannot guarantee one-way delay faster than the speed of light on any long-haul network. The need for QoS arises when there are multiple data streams competing for the limited physical capacity of the transmission media or network devices (see Figure 4.1). In the case of WiMAX, the limiting resource is the *radio frequency bandwidth*. When there are multiple data streams competing to use the same frequency bandwidth, a QoS policy is needed to determine which data stream has the priority to use the air interface. This QoS policy depends on the user applications that are characterized by QoS performance metrics. For example, an e-mail application does not need any guarantee except for reliable delivery of the data. A VoIP application needs guarantee of low latency. A video-streaming application can afford a long delay but requires relatively high bandwidth. The following elements are required to implement QoS on a network:

1. *QoS performance metrics*: QoS is a mechanism to assure network performance as defined by a set of metrics associated with each data stream. Examples of performance metrics are delay, throughput, jitter, and packet loss.

2. *Request and grant*: This is also known as *admission control*. In the case of WiMAX, the BS is the central control point. An SS requests a connection with certain QoS parameters. If the network does not have the resource, the request will be rejected. If the network has sufficient resource, the BS will check if the SS is authorized to use the resource. After authorization, the BS will guarantee the service throughout the connection.

3. *Traffic shaping*: For an incoming packet, the network device needs to determine how to classify the packet and whether to send the packet. If the packet delivery is not guaranteed and the network is congested, the packet could be dropped. Otherwise, the packet enters into a priority queue and waits for the scheduler to determine its delivery. The IEEE 802.16 standard does not require traffic shaping as the air interface would not drop packets. The traffic

shaping on the wireline side of the device is outside the scope of the standard.

4. *Scheduling policy*: A QoS-enabled device has multiple priority queues for different classes of services. The scheduling policy is to determine how and when to process packets in the priority queues. A scheduling policy could use a round-robin method to process packets in each priority queue and allocate more resources for high-priority queues. Another scheduling policy could be to process packets in a low-priority queue only when all high-priority queues are empty.

4.2.2 QoS in Circuit-Switching Network

In a circuit-switched network, each data stream has a dedicated connection (called circuit) from a source to a destination. As there is no competition from multiple data streams, the performance is determined by (a) the network device, (b) the transmission media, and (c) the distance of the transmission. There is no need for QoS in a circuit-switched network because the service is guaranteed due to the *deterministic* nature of the network. An example is the traditional public switched telephone network (PSTN). If a packet-switching technology emulates the circuit-switched network, known as circuit emulation service (CES), its performance would be guaranteed due to the nature of a circuit-switched network.

4.2.3 QoS of Packet-Switched Network

On a wired network, congestion occurs on the device and never happens on the media (i.e., the wired cable). For a connectionless service, the QoS information is carried on the packet itself. For a connection-oriented service, the QoS is configured on the device.

4.2.3.1 Connectionless Service (Ethernet or IP)

In a *connectionless* network (Ethernet or IP), each packet needs to carry the QoS information in its packet header. For Ethernet, the IEEE 802.1p standard supports a 3-bit priority scheme and it allows up to eight priority classes [4] while most implementations support only two queues (priority $= 0$ and priority $\neq 0$). An incoming Ethernet frame is put into one of the queues. Frames in the high-priority queue are processed first. The Internet Protocol (IP) also allows IP packets to carry QoS information in the packet header, which is a 6-bit type of service (ToS) field. An IP-router with QoS capability has multiple queues for incoming traffic based on the ToS field. In addition, there are various protocols to support end-to-end QoS on an IP network [3].

4.2.3.2 Asynchronous Transfer Mode

Asynchronous transfer mode (ATM) is a *connection-oriented* service. Each connection is called an ATM virtual circuit that inherits many essential features of a circuit-switched network. The QoS parameters are specified for each virtual circuit rather than on individual ATM cells. The ATM adaptation layer (AAL) specifies the following QoS schemes for ATM virtual circuits:

1. Constant bit rate (CBR): A CBR connection has a guaranteed bandwidth and it supports CES that is to emulate DS0 or T1 circuit. There is no need for QoS of a CBR connection as the services are guaranteed by the nature of circuit-switched technology. Many carriers are using ATM CBR to carry voice traffic on their ATM backbone.

2. Variable bit rate (VBR): This is the most flexible QoS scheme. A VBR connection has three parameters—peak cell rate (PCR), sustained cell rate (SCR), and maximum burst size (MBS). A VBR connection guarantees the service at the SCR level. If there is available bandwidth after SCR, it would burst the traffic to the PCR level. For example, if a customer subscribes to the service of SCR = 512 Kbps and PCR = 1024 Kbps, the customer is guaranteed to have a data rate up to 512 Kbps. If the network is not busy, the customer would enjoy up to 1024 Kbps. The VBR service includes real-time VBR (rtVBR) and nonreal-time VBR (nrtVBR), where rtVBR provides assurance of delay for real-time applications such as video conference.

3. Unspecified bit rate (UBR): This is also known as the best effort service. In other words, there is no guarantee of this service. If the network has a bandwidth, it will serve UBR connections. If the network is congested, the UBR cells would be put on a waiting queue. When the queue buffer is full, the cells are dropped.

There are more AAL services (such as ABR), but this chapter covers only four—CBR, rtVBR, nrtVBR, and UBR—as they are related to the WiMAX QoS services to be discussed later. An ATM switch with QoS capability supports admission control that is configured via either permanent virtual circuit (PVC) or switch virtual circuit (SVC). PVC is a manual provision and SVC uses the Q.2931 protocol to establish a virtual circuit. If an ATM switch does not have the capacity to support a connection request, it rejects this request. If an ATM switch accepts a connection request, it guarantees the QoS until the connection is terminated manually or via the Q.2931 protocol.

4.2.4   QoS of Wireless LAN and 802.11e

Like Ethernet, WLAN (802.11) is a *connectionless* service. However, the bottleneck of WLAN is the *transmission media* (i.e., the radio frequency spectrum) rather than the network device. As a result, the QoS schemes discussed for Ethernet, IP, and ATM do not apply to WLAN. The 802.11 standard supports two access methods—distributed coordination function (DCF) and point coordination function (PCF) [5]. The PCF operation requires the wireless access point (WAP) to function as a central control point that polls each wireless client at a regular interval. A wireless client is allowed for data transmission only when it is *polled*. The WAP could implement a QoS scheme that polls real-time applications more often than nonreal-time applications. Although PCF has the capability to support a QoS scheme, PCF is not supported by most vendors. As a result, a new QoS scheme is proposed by the IEEE 802.11e working group (which is still a draft). 802.11e specifies a new access method, extended DCF (EDCF), which uses different interframe gap (IFG) and contention windows (CW) for differentiated services. A real-time service will have a shorter CW and IFG while a nonreal-time service will have a longer CW and IFG. As a result, a real-time application will have a higher probability of accessing the media. Experimental results of EDCF are mixed [6] and few vendors are supporting the draft standard, yet.

4.3   IEEE 802.16

4.3.1   Basic Operation

The IEEE 802.16 standard supports two network architectures—P2MP mode and mesh mode. The mesh mode is an optional architecture and the discussion of QoS performance and scheduling of the mesh mode can be found in Ref. 7. This chapter focuses on the QoS performance and scheduling of the P2MP mode and its network architecture is illustrated in Figure 4.2. The

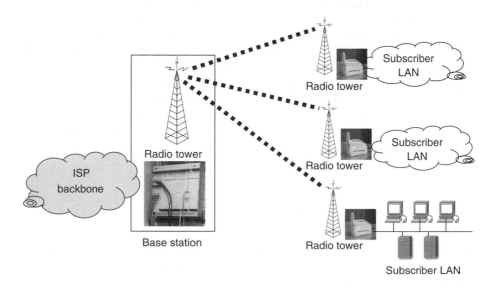

FIGURE 4.2
WiMAX P2MP network.

P2MP network has one BS and multiple SSs. The BS is the central control point and regulates all the traffic on the network. IEEE 802.16 is a connection-oriented service with which each SS needs to establish a service connection to the BS. An SS first sends a message to the BS for requesting services on the network. The connection between the BS and an SS could be either DL (from BS to SS) or UL (from SS to BS). The protocol stack of WiMAX is illustrated in Figure 4.3 and the connection-oriented service is defined in the MAC sublayer.

The multiple access schemes in WiMAX include both frequency division duplex (FDD) and time division duplex (TDD). The QoS discussion in this chapter is on TDD only because TDD is more flexible than FDD for QoS implementation. Within a given frequency bandwidth, IEEE 802.16 supports an adaptive scheme to allocate time slots on UL and DL channels as illustrated in Figure 4.4. Each frame is broken into multiple time slots, and the BS can dynamically allocate different time slots for DL and UL. In the case of Internet application where the data transfer is mostly from the Internet down to a subscriber, the BS allocates more time slots for DL and fewer for UL. For VoIP applications, the BS allocates the same number of time slots for DL and UL.

Upper layer			
Service specific convergence sublayer			
MAC sublayer common part			
Security sublayer			
Transmission sublayer			
QPSK	QAM-16	QAM-64	QAM-256

FIGURE 4.3
WiMAX protocol stacks.

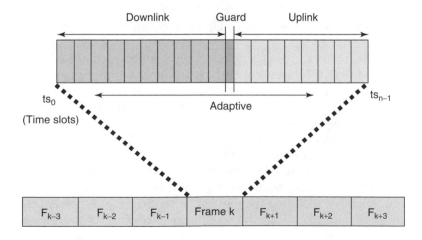

FIGURE 4.4
Time division duplex (TDD) framing.

4.3.2   Service Flow

When the BS receives a connection request from an SS, it calls for an authorization module to determine if the user has the proper authorization for the service. The BS then determines if the physical resource (i.e., RF bandwidth) is available to support the request. If yes, the BS associates the connection request with a *service flow* with the requested QoS parameters. Each connection is identified by a 16-bit connection ID (CID). Note that IEEE 802.16 does not use source or destination MAC addresses in the MAC frame. A service flow has the following attributes:

a. Service flow ID (SFID): Each service flow has an SFID with its transmission direction (DL or UL).
b. CID: A CID is mapped to an SFID after the connection is admitted.
c. Provisioned QoS parameters: QoS is provisioned via a network management system.
d. Admitted QoS parameters: QoS parameters for which the BS is reserving the resources. The primary resource to be reserved is the bandwidth.
e. Active QoS parameters: QoS parameters actually provided for the service flow. Only active service flow may send packets over the wireless link.

A service flow could be *statically* provisioned through the network management system or dynamically created by the following IEEE 802.16 control messages:

- Dynamic service addition (DSA): to create a new service flow
- Dynamic service change (DSC): to change an existing service flow
- Dynamic service deletion (DSD): to delete an existing service flow

These MAC control messages allow a service provider to add new subscribers, modify QoS for existing customers, allocate more resource (i.e., RF bandwidth) to existing links, and reclaim unused resources. All this can be accomplished during the operation without interfering with the active services of existing customers.

A dynamic service request can be initiated by either the BS or an SS. In the case of SS-initiated request, a DSA-REQ (request) message is sent from an SS with a service flow reference and the QoS parameter set. When the BS receives the DSA-REQ message, it sends a DSX-RVD (received) message to inform the SS of receiving the request. After that, the BS sends DSA-RSP (response) message to indicate the acceptance or rejection of the request. The SS then sends an acknowledgment (DSA-ACK) . The complete message flow diagram is illustrated in Figure 4.5. The BS-initiated request is similar except that there is no need to send a DSX-RVD.

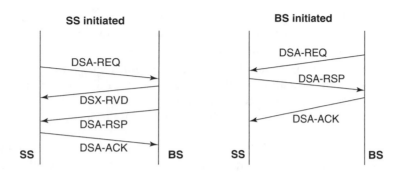

FIGURE 4.5
Dynamic service addition (DSA)—message flow.

The DL transmission is controlled by the BS and MAC frames are broadcasted to all SSs. When an SS receives a frame, it checks for the CID in the frame. To improve the transmission performance, multiple frames are combined into a *burst*, which is then sent to the air interface for broadcasting. When an SS receives the burst, it will retrieve only the frames with its own CID and discard the other frames. Each burst includes a DL-MAP and UL-MAP, which specify the structure of the burst and how to retrieve the MAC frames from the burst. The UL data transmission is more complex as all SSs must synchronize with the BS for data transmission. The details of the UL data transmission are related to the QoS scheme discussed in Section 4.3.3.

4.3.3   QoS in IEEE 802.16

The IEEE 802.16 standard specifies the following four classes of services [8,9]:

1. Unsolicited grant service (UGS): This service is for transmitting uncompressed voice and to emulate circuit-switched services such as DS0, $n \times$ DS0, and T1. This service requires a fixed amount of data at a fixed time interval and guarantees the throughput and delay for the service.

2. Real-time polling service (rtPS): This service is for compressed multimedia (such as video streaming) and other real-time applications where the amount of bandwidth requirement may vary at each instant. This service requires the BS to implement a *polling* mechanism to the SSs at a fixed interval. Each poll asks the SS to specify the bandwidth requirement for each time interval. The polling is on the DL channel to avoid contention by the SSs.

3. Nonreal-time polling service (nrtPS): This service is for nonreal-time application that requires a guaranteed performance. It requires the BS to poll the SSs at a fixed time interval, but not at a rigid time interval as rtPS. If an SS does not respond to the poll after n times in a row, the BS will put the SS in a waiting group.

When the waiting group is polled, all SSs in the group will be contending for network access. This mechanism prevents stations with little traffic to waste valuable polls.

4. Best effort (BE) service: This service does not require a poll. An SS must contend with other SSs for bandwidth and network access. Requests for bandwidth are in the time slots marked in the UL-MAP as available for contention. If a request is successful, it will be indicated in the next DL-MAP and the SS can transmit the data. If it is not successful, the SS must try again later. It is possible to have collisions for the request, and the same back-off algorithm for Ethernet is applied when collision occurs.

The following is the set of performance metrics to support the above QoS classes and the application of these metrics to specify the QoS schemes is given in Table 4.1:

- *Maximum sustained traffic rate (MSTR)*: The peak data rate (in bps) of a service flow. This parameter is comparable to ATM-PCR. This service rate shall be policed for the wireless link to assure its conformance as measured in average over time.

- *Minimum reserved traffic rate (MRTR)*: The minimum reserved data rate (in bps) for a service flow. This rate is guaranteed for the service. This parameter is comparable to ATM-SCR. For example, a user subscribes to a service of MRTR of 768 Kbps with an MSTR doubling the rate to 1.544 Mbps. The user is guaranteed 768 Kbps for throughput, plus other contractual service of latency and jitter. If the service provider network has a bandwidth available beyond MRTR, the service provider would allow the user traffic to continue up to the specified MSTR. The user shall not send the data beyond the MSTR level.

- *Maximum traffic burst*: The maximum burst size (in bytes) for a service flow. This parameter is comparable to ATM-MBS.

- *Maximum latency*: The maximum latency (in milliseconds) between the reception of a packet by BS/SS and the transmission of the packet by SS/BS.

TABLE 4.1

QoS Classes and Parameters

	UGS	rtPS	nrtPS	BE
Minimum reserved traffic rate		✓	✓	
Maximum sustained traffic rate	✓	✓	✓	✓
Maximum traffic burst		✓	✓	
Tolerated jitter	✓			
Maximum latency	✓	✓		

- *Tolerated jitter*: The maximum delay variation (in milliseconds) of a service flow.

4.3.4 Admission Control

A connection request could be initiated by either the BS or an SS as illustrated in Figure 4.5. IEEE 802.16 uses a mechanism of *request* and *grant* for connection-oriented services. A request is from an SS to inform the BS that it needs bandwidth. When an SS sends a request to the BS for a connection with certain QoS parameters, the BS first authenticates the SS. After authentication, the BS needs to determine if the resource is available for the request:

$$\sum (\text{all committed bandwidth}) + \text{new bandwidth request} \leq \text{total bandwidth}$$

In the case of UGS, the committed bandwidth is MSTR while in the case of rtPS and nrtPS the committed bandwidth is MRTR. The BE service has no committed bandwidth. An important note about IEEE 802.16 is that the UL and DL bandwidth could be dynamically allocated based on the user needs. If there are more demands for DL while many UL time slots are available, the BS can allocate more time slots from the UL to the DL.

A bandwidth request message is usually transmitted during a UL allocation (SS => BS), and the standard also allows an optional provision for piggyback request. It should be noted that the request is sent in the contention mode and could be lost (due to collision). As a result, the BS needs to issue the message DSX-RVD to confirm the reception of the DSA-REQ message. The BS issues the bandwidth grant in the UL-MAP that is broadcasted to all SSs, and individual SSs use CID and the UL-MAP to retrieve its own grant. After a connection is created with the service flow parameters, an incoming packet for the transmission enters into a priority queue to be served by the QoS scheduler.

4.3.5 QoS Scheduling

When a connection request is granted, a service flow with the QoS parameters is created for the connection. Scheduling services is the data-handling mechanism to support the MAC scheduler for data transport on a connection. The BS controls both the UL and DL scheduling as illustrated in Figure 4.6, and this approach is similar to the QoS architecture in Ref. 10. The scheduler calculates the throughput and latency requirements of the UL and DL traffic and provides the *polls* and *grants* at the appropriate time intervals. The DL is broadcast and the scheduler fills in each *burst* based on the QoS parameters of the frames in the queue. The UL scheduling uses a poll/grant scheme that is more complex as it requires coordination between the BS and individual SSs.

4.3.5.1 UGS Scheduling

The UGS is designed to support real-time service flow of *fixed*-size data packets on a fixed interval. The services provide fixed-size grants on a regular

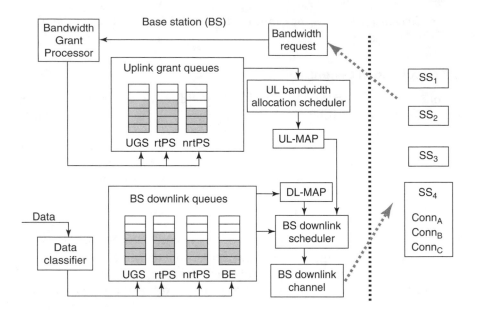

FIGURE 4.6
Base station UP and DL scheduling.

basis and eliminate the latency of SS request to assure that the real-time needs are met. The BS provides grant to the SS at a fixed interval based on the MSTR (see Table 4.1) and the size of the grants is large enough to hold the fixed-length data plus the MAC overhead. The SS receives broadcast frames from the BS at regular intervals, and the UL-MAP in the broadcast frame contains the UL channel for the SS to send the data. When an SS has the data to send, it puts the data into the assigned channel (time slot) and bursts the data to the air interface immediately. There is no bandwidth sharing of multiple connections for the UGS service and each connection (service flow) is allocated with a dedicated channel (time slot) for the UL data transmission.

4.3.5.2 rtPS Scheduling

The rtPS is designed to support real-time service with *variable*-size data packets on a fixed interval, such as streaming audio and video. The service allows an SS to specify the size of the desired grant and it has more request overhead than UGS. The BS issues request opportunities for the SSs to obtain UL transmission opportunity. Multiple connections of rtPS share the same bandwidth for the UL data transmission and a connection can send the data on the UL channel only when it is polled. The implementation of the polling service is not specified in the standard and each vendor may design its own polling mechanism. The following is an example of the polling service:

1. An SS requests and is granted a connection of the rtPS service with a guaranteed bandwidth of 378 Kbps and a delay of 50 ms.

2. The BS shall poll the SS at a fixed interval shorter than 50 ms.

3. Whenever the SS has data to send it first waits for its polling, which is sent from the BS to the SS on the broadcast DL channel. The SS checks for its own polling as indicated in the UL-MAP with its own CID.

4. When the SS gets its polling period, the SS retrieves the bandwidth allocation information from UL-MAP and uses the bandwidth to send the data on the assigned UL channel.

5. If there is more data to send than the allocation, the SS shall build the frame according to its guaranteed bandwidth of 378 Kbps and maximum traffic burst.

6. After sending the data burst to the air interface, the SS waits for its next poll.

4.3.5.3 nrtPS Scheduling

The nrtPS provides polls on a regular basis and assures that the service flow receives request opportunity even under network congestion. In general, the BS polls nrtPS connections at an interval of 1s or less. The BS shall provide the request opportunities to SSs as specified by the QoS parameters. In addition, the SSs are allowed to use contention request opportunities to obtain grants.

4.3.5.4 BE Scheduling

The BE service is to provide an efficient operation for best effort traffic. The SSs are allowed to use contention request opportunities to obtain grants. Collision could happen when multiple stations are transmitting requests at the same time. When collision happens, each SS uses a back-off algorithm similar to 802.11, except that the contention window is controlled by the BS, which uses the DL channel to specify the contention window size for individual SSs. The grants to SSs are sent via the DL channel, which uses the UL-MAP to specify the channel for UL data transmission. Note that the BS does not have a scheduler for the BE requests as they are operating in the contention mode. The BS scheduler, however, handles BE requests and provides grants (which are UL channels) for UL data transmission. There is no contention for UL data transmission.

4.4 Simulation of IEEE 802.16 QoS Operation

4.4.1 Admission Control

The policy of admission control is similar to the Erlang B model that has been used by the voice networks for many years [11]. We apply the same concept of the Erlang B model and use the IEEE 802.16 QoS parameters to describe the model behaviors. Since the UL and DL use separate channels, the admission control is applied to DL or UL separately. The procedure of admission control is required for UGS, rtPS, and nrtPS. BE requests are always granted as there is no committed bandwidth. In the simulation model of admission control, we have the following parameters:

1. Number of SSs: this is a fixed parameter during the simulation.
2. Connection requests per minute per SS (λ): This is the arrival rate to or from each SS and it follows the Poisson distribution. During the simulation, we use *the interarrival rate* (μ) to determine the time interval of the next request

$$\text{Interarrival rate } (\mu) = 1/\lambda.$$

   This parameter (μ) follows the exponential distribution in the simulation mode.
3. Bandwidth request (in multiples of DS0): This is either MSTR for UGS or MRTR for rtPS/nrtPS. It follows the exponential distribution. The bandwidth request shall be at least one DS0 (64 Kbps); otherwise, no data can be transmitted. The average bandwidth request is set at $4 \times 64 = 256$ Kbps.
4. Data size (S) in bytes: This parameter also follows the exponential distribution. The data size and bandwidth request determine the service time of a request after its admission. For example, if the data size is 2 Mb and the requested bandwidth is 1 Mbps, the duration of the service will be $2M \div 1.0M \times 8 = 16$ s. This is based on the store-and-forward scheme used in most network devices. The average data size is set at 2 MB.
5. Total bandwidth (T) in bps: This is the total bandwidth allocated to either a DL or UL channel. As discussed earlier, bandwidth request could be provisioned manually or dynamically allocated. If it is statistically provisioned, the subscriber does not need to request for admission. Therefore, the simulation model of admission control is for dynamic service requests only. The total bandwidth is fixed at 27 Mbps for the simulation.

The first simulation is to study the blocking probability (the percentage of rejected requests) and its relation with the number of SSs and the request rate (λ). Each simulation run lasts for 10–20 min. The results of blocking probability versus SSs with three different arrival rates ($\lambda = 1$, 2, and 4) are illustrated in Figure 4.7. The second simulation is to perform a sensitivity analysis of blocking probability and requested bandwidth with $\lambda = 2$ and SS = 50, and the result is illustrated in Figure 4.8. The third simulation is to measure the channel utilization versus SSs (SS = 20, 50, 100, and 200) with fixed parameters of bandwidth = 2 and $\lambda = 1$ for a 20-min simulation run. The results are illustrated in Figure 4.9.

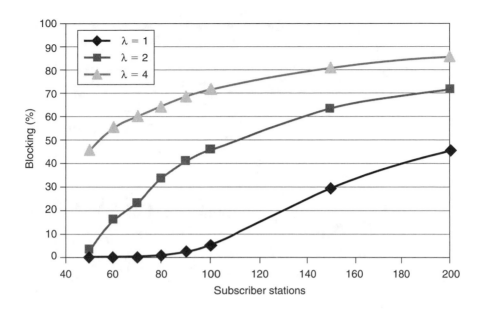

FIGURE 4.7
Blocking probability versus number of subscriber stations.

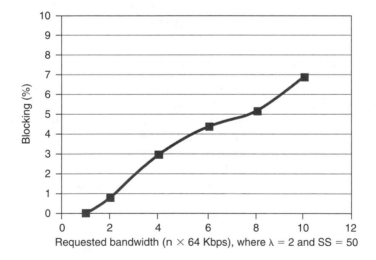

FIGURE 4.8
Sensitivity analysis on requested bandwidth.

These simulation results provide a guideline for engineering the QoS services for the WiMAX subscribers. For example, a network engineer may allocate only 50% of the bandwidth for MSTR of UGS and MRTR of rtPS/nrtPS, so that the network can have sufficient capacity to serve BE subscribers. If the arrival rate for UGS and rtPS/nrtPS is one request per minute

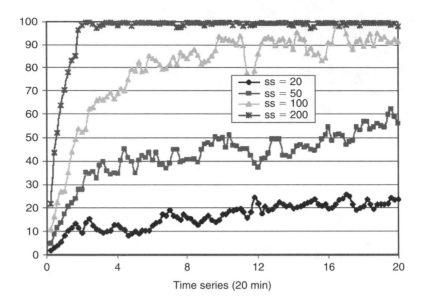

FIGURE 4.9
Channel utilization of UGS-only traffic.

($\lambda = 1$) with average data size of 2 MB, the engineer guideline is to support up to 50 subscribers. If the engineering rule changes to 70% for UGS and rtPS/nrtPS subscribers, the network could support up to 75 subscribers.

4.4.2 Simulation of IEEE 802.16 QoS Scheduling

The UL and DL bandwidths are controlled by the BS. In the case of DL data transmission, the frames are combined into *burst* and broadcasted to all SSs. In the case of UL data transmission, an SS waits for its polling interval and transmits the data only when it is polled. In the simulation runs, we exclude the manual provision of UGS as it is similar to a circuit-switched service. The simulation is based on the dynamic bandwidth requests as illustrated in Figure 4.5. After the request/grant process for admission, a user request is entered into the scheduling queues of UGS, rtPS, nrtPS, and BE. The scheduler uses the poll/grant mechanism to determine how to serve the jobs in the queues. In the simulation, we use two sampling intervals—10 ms (for UGS and rtPS) and 50 ms (for nrtPS and BE). The configuration of simulation profile is given as follows:

- DL/UL bandwidth = 14 Mbps.
- Basic channel = 64 Kbps (DS0).
- Number of DL/UL channels: 14 M ÷ 64 K = 218 channels.
- Number of SSs = 100.
- Request arrival rates per station (λ): 2/min (light traffic) and 5/min (heavy load).

TABLE 4.2

QoS Scheduling Simulation

	Light Traffic ($\lambda = 2$)		Heavy Traffic ($\lambda = 5$)	
	Comp. Jobs	Avg. bandwidth (bps)	Comp. Jobs	Avg. bandwidth (bps)
UGS	469	512	1164	512
rtPS	471	512	1197	510
nrtPS	505	512	1264	453
BE	494	512	677	15

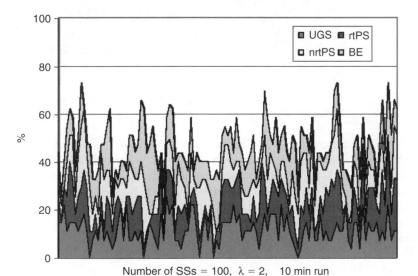

Number of SSs = 100, $\lambda = 2$, 10 min run

FIGURE 4.10
Simulation of channel utilization (light traffic).

- Request profiles: equally distributed for UGS (25%), rtPS (25%), nrtPS(25%), and BE (25%).
- Bandwidth per request: 8 × 64 Kbps. For UGS it is 100% MSTR. For rtPS/nrtPS it is 50% MSTR and 50% MRTR. For BE it is all MSTR. We ran the simulation for both fixed bandwidth and exponentially distributed bandwidth. The results presented are based on *fixed* bandwidth for ease of analysis.
- Average data size per request: 250 KB (exponential distribution).
- Simulation run: 10–20 min.

The results of the simulation are illustrated in Table 4.2, Figure 4.10 (light traffic), and Figure 4.11 (heavy traffic). We observe that during the light traffic load, each traffic class receives almost the same bandwidth allocation. During

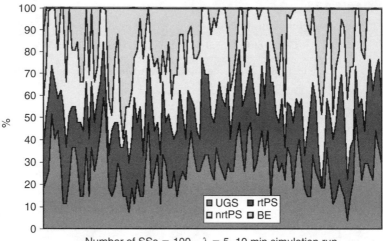

Number of SSs = 100, $\lambda = 5$, 10 min simulation run

FIGURE 4.11
Simulation of channel utilization (heavy traffic).

the heavy load, only the UGS class is able to maintain the same performance level. For rtPS and nrtPS, they maintain the level of MRTR. The BE class receives very little bandwidth allocation from the BS. This simulation results conform to the expected behavior of WiMAX QoS, and are consistent with other studies in the literature [12,13].

4.5   Conclusions

This chapter presents the concept and requirements of QoS as specified in the IEEE 802.16 standard, along with an architecture to implement QoS in a simulation model. As presented in this chapter, the QoS requirements specified in IEEE 802.16 are similar to ATM QoS, and the QoS procedure is based on poll/grant that is similar to PCF of 802.11. The support of QoS is essential for BWA because service providers can use it to offer differentiated services. The IEEE 802.16 standard does not provide the details of admission control and QoS scheduling, and this chapter fills the gap to implement a solution for it. Another contribution of this chapter is the development of a simulation model. The results from simulation conform to the expected behavior of QoS as specified in IEEE 802.16 and are consistent with other studies.

4.6   Acknowledgments

This research project is partially supported by the Quality Instruction Council (QIC) grant of DePaul University. The author would like to thank ISP, Inc.

at British Columbia, Canada for its generous donation of a high-capability Linux server that is used for the simulation of this project.

References

1. IEEE 802.16-2004, *Air Interface for Fixed Broadband Wireless Access System*, 2004.
2. *Business Communications Review*, p. 6, August 2006.
3. A. S. Tanenbaum, *Computer Networks*, Prentice-Hall, Upper Saddle River, New Jersey, pp. 397–417, 2003.
4. IEEE 802.1D, *Medium Access Control (MAC) Bridges—including 802.1p and 802.1w*, 1998.
5. IEEE 802.11, *Wireless LAN Medium Access Control (MAC) and Physical Layer*, 1999.
6. H. M. Liang, C. H. Ke, C. K. Shieh, W. S. Hwang, N. K. Chilamkurti, *Performance Evaluation of 802.11e EDCF in Infrastructure Mode with Real Audio/Video Traffic*, International Conference on Networks and Services, p. 92, July 2006.
7. M. Cao, W. Ma, Q. Zhang, X. Wang, W. Zhu, *Modeling and Performance Analysis of the Distributed Scheduler in IEEE 802.16 Mesh Node*, MobiHoc'05, pp. 78–89, May 25–27, 2005.
8. M. Pidutti, *802.16 Tackles Broadband Wireless QoS Issues*, http://www.commsdesign.com, ArticleID=54201623, December 2004.
9. B. Hayat, R. M. A. Nasir, *802.16-2001 MAC Layer QoS*, http://www.acm.org/ubiquity/views/v7i17_hayat.html.
10. G. Chu, D. Wang, and S. Mei, *A QoS Architecture for the MAC Protocol of IEEE 802.16 BWA System*, IEEE International Conference on Communications Circuits, vol. 1, pp. 453–439, 2002.
11. H. Wang, B. He, D. P. Agrawal, *Admission Control and Bandwidth Allocation above Packet Level for IEE 802.16 Wireless MAC*, 12th International Conference on Parallel and Distributed Systems, 2006.
12. D.-H. Cho, J.-H. Song, M.-S. Kim, and K.-J. Han, *Performance Analysis of the IEEE 802.16 Wireless Metropolitan Area Network*, 1st International Conference on Distributed Frameworks for Multimedia Applications, 2005.
13. P. Neves, S. Sargento, R. L. Aguiar, *Support of Real-Time Services over Integrated 802.16 Metropolitan and Local Area Networks*, 11th IEEE Symposium on Computers and Communications, June 2006.

5

Propagation and Performance

Thomas Schwengler

CONTENTS

Modern wireless communication systems deliver reliable high-speed data services. Consumer expectations have become very high: cheaper rates, higher data throughput, flexible applications, better service integration, and almost ubiquitous availability are expected from wireless service providers. Radio technologies and standards have been successful in delivering many of these expectations: CDMA-based third-generation systems such as EV-DO (IS-858) and HSDPA provide affordable multimegabit services, and are available in major cities. IEEE standard 802.16 and the WiMAX Forum are pursuing similar goals and present another high-speed access alternative. WiMAX offers both fixed and mobile systems, efficient and adaptive coding and modulation techniques, scalable channel sizes, subchannelization schemes, MIMO antenna systems, quality of service (QoS), and more.

High-speed wireless services have already achieved great success in local area networks (LAN) with the IEEE 802.11 standard and Wi-Fi certified products. The goal is now to broaden wireless access to metropolitan area networks (MAN) and complement current wired services such as ADSL and cable modem.

This chapter presents carriers' perspectives for wireless services like fixed WiMAX access. Of course, fundamentals of wave propagation are still of the utmost importance, and the nature of wireless channels (including their relative unpredictability and fading characteristics) must be well understood. Before deploying new wireless services on a large scale, service providers need a good estimate of capacity and coverage of these systems. To this end, this chapter presents various aspects of propagation and performance for WiMAX radio systems: it reviews WiMAX radio system parameters such as link budgets; it presents relevant propagation models; and it analyzes system throughput and performance for a typical suburban area.

5.1 Introduction

IEEE 802.16 is a standard for wide area wirelss networks. It includes important service providers requirements such as QoS, security, flexible and scalable operations in different RF bands. WiMAX goes one step further and narrows down some implementation choices of 802.16 to achieve interoperation between equipment manufacturers. WiMAX standardizes several air interfaces and several profiles in different frequency bands. Of course, performance varies with frequency, channel bandwidth, and other profile characteristics; and conformance between products and suppliers exist only in a given profile.

5.1.1 Fixed and Mobile

Two very different families of WiMAX systems exist and should be treated separately: fixed and mobile WiMAX. In addition, a regional initiative, WiBro, which resembles mobile WiMAX, has been standardized in Korea.

Fixed WiMAX is a reliable and efficient air interface, based on 802.16-2004 [1], used for fixed broadband access. Several profiles exist for fixed WiMAX, including different bandwidths, carrier frequencies, and duplexing schemes: time division duplexing (TDD) and frequency division duplexing (FDD). Its air interface is based on orthogonal frequency division multiplexing (OFDM) and access between multiple users within a sector is managed by time-division multiple access (TDMA). While equipment has been available since 2004, major milestones were achieved in 2005 when suppliers demonstrated successful intervendor operations. Conformance testing [2] led to the first WiMAX equipments to be certified in January 2006.

Fixed WiMAX profiles at 3.5 MHz (TDD and FDD) in the 3.5 GHz band were the first to be certified and will be examined in this chapter; 10 MHz TDD channels at 5.8 GHz are another important profile and will also be studied.

Mobile WiMAX is an extension of the above that includes a new standard for mobility: 802.16e-2005 [3]. Mobile operations require more complexity in the air interface and in the network architecture. Therefore, mobile WiMAX defines a different standard with considerations such as location register, paging, handoff, battery-saving modes, and other network functions to manage mobility. Its air interface is based on orthogonal frequency division multiple access (OFDMA).

Release-1 Mobile WiMAX profiles cover 5, 7, 8.75, and 10 MHz channel bandwidths for operations in the 2.3, 2.5, 3.3, and 3.5 GHz frequency bands. Plugfests showing interoperability between suppliers started in September 2006.

WiBro is a Korean initiative for wireless broadband. Similar in many ways to mobile WiMAX, WiBro includes mobility and handoff, and is commercially available in Korea since mid-2006.

WiBro operates in 10 MHz TDD channels at 2.3 GHz and uses OFDMA. It targets mobile usage up to 60 mph.

The standard community is now almost exclusively focusing on mobile WiMAX, for both air interface and end-to-end network architecture [4,5]. Still, fixed WiMAX applications should not be overlooked; small and large service providers have conducted over 100 major fixed WiMAX trials. This precious experience, combined with mobile cellular data expertise, give us a wealth of information to better design future broadband access services.

5.1.2   Frequency

WiMAX is a flexible and scalable standard that may be adapted to different frequency bands. The standard is torn between two opposite goals. On the one hand, limiting frequency bands and channel bandwidths narrow down the standard and make interoperability easier while on the other, profiles in different bands and using different channel widths make the standard more flexible.

Frequency bands and frequency channel widths are standardized in different WiMAX profiles. There are many reasons behind the choices made for these bands, including spectrum availability and regulations in different countries. The bands of highest interest for WiMAX are presented below:

2.3 GHz: In the United States, a 1997 auction for wireless communications service (WCS) addressed 30 MHz of spectrum, which was then left unused for a long time. WiBro-related products may soon change that.

2.5 GHz: Educational broadband services (EBS) and broadband radio services (BRS) occupy a large band of spectrum above 2.5 GHz.* Renewed interest comes from the high priority given to these bands for 802.16e mobile WiMAX products.

Broadband access at 3.4–3.7 GHz: In many countries, the spectrum between 3.4 and 3.6 GHz was allocated (in most cases auctioned) for fixed broadband wireless access. This band was the first to see WiMAX certified products. In the United States, 3.65–3.7 GHz was allocated in March 2005 for fixed and mobile service, which unfortunately provides much less spectrum. Operations in the band should be licensed on a nationwide nonexclusive basis with all licensees registering their fixed stations in a common database.† Protection zones of 150 km were established around the grandfathered fixed satellite stations.

Unlicensed spectrum at 5.4–5.8 GHz: In the WiMAX community, some equipment manufacturers and service providers are interested in unlicensed (or license exempt) bands of spectrum. In the United States, these bands are governed by Part 15 of the FCC Rules & Regulations: they may not cause harmful interference to authorized services and have to follow listen-before-talk rules.

Several unlicensed bands exist and have great potential for fixed access, but only the highest is the focus of WiMAX. There are several reasons for this: the 900 MHz band benefits from great propagation characteristics but is limited in power and bandwidth; the 2.4 GHz band is wider but has recently seen heavy deployment of Wi-Fi LANs.

The 5 GHz band is referred to as the unlicensed national information and infrastructure (UNII) band. Its upper portion (UNII-3, 5.725–5.825 GHz) is intended for community networking communications devices operating over a range of several kilometers.

---

* Formerly MMDS and ITFS, these spectrum bands are now referred to as EBS and BRS spectrum bands. A new band plan was proposed by FCC to transition the old 6 MHz analog TV channels to 5.5 MHz channels.

† The WiMAX Forum and several member companies have asked the FCC to adopt an exclusive licensing regime for the 3.65–3.7 GHz band in the top 50 metropolitan statistical areas (MSAs), while retaining its nonexclusive licensed approach in smaller markets.

Combined with a new 5.475–5.725 GHz* band recently opened by the FCC, over 400 MHz of spectrum is now available for unlicensed operations.

Other bands of spectrum are of interest to the WiMAX Forum, although no specific profiles have been defined for them yet.

UHF channels at 700 MHz: TV broadcasting spectrum is very attractive for broadband wireless applications because of its excellent propagation and in-building penetration characteristics. In the United States, TV broadcasters must transition to digital television and return their 700 MHz analog frequencies by February 18, 2009. This opens large bands of spectrum for potential use in wireless communications. Suppliers are already developing equipment in these bands based on 802.16 and WiMAX.

AWS at 1.7–2.1 GHz: In August 2006, the FCC auctioned 90 MHz of spectrum for advanced wireless services (AWS). This band was somewhat puzzling to equipment manufacturers because of its pairing with a rather large interval between forward and reverse links (400 MHz); still, WiMAX and 3G services can be expected in this band.

Public safety at 4.9 GHz: In 2002, the FCC designated 50 MHz of spectrum in the 4.9 GHz band for exclusive public safety use. WiMAX services are appropriate for public safety applications. Products exist in that band and plugfest initiatives started in 2006 for operations between suppliers.

5.2 Propagation Environment

Propagation environments are certainly not specific to WiMAX, but WiMAX performance levels in different environments should be quantified. Propagation characteristics depend on the bands of operations and are reviewed in this section.

5.2.1 Propagation Modeling

Different spectrum bands have very different propagation characteristics and require different prediction models. Some propagation models are well-suited for computer simulation in the presence of detailed terrain and building data; others aim at providing simpler general path loss estimates [6].

A handful of empirical models were widely accepted for cellular communications; their success being mostly due to their simplicity and their fairly good

---

* Rules are similar to UNII-3, but with requirements around dynamic frequency selection (DFS) capability to protect Federal Government radar systems.

prediction for first-order modeling. The simplest approach is to estimate the power ratio between transmitter and receiver as a function of the separation distance d, that ratio is referred to as path loss. A physical argument like the Friis' power transmission formula yields:

$$\frac{P_r}{P_t} = \frac{G_t G_r \lambda^2}{(4\pi d)^2} \tag{5.1}$$

where P_t and P_r are the transmitted and received power, G_t and G_r the transmitter and receiver gain, λ the wavelength of the signal, and d the separation distance. This equation shows a free-space dependence in $1/d^2$. The exponent $n = 2$ is referred to as the path loss exponent. If the path loss is measured in decibel ($PL = 10 \times \log(P_t/P_r)$), it varies logarithmically with the distance of separation. Simple models then consist of computing a path loss exponent n from some linear regression argument on a set of field data, and deriving a model like:

$$PL(\text{dB}) = PL_0 + 10n \times \log(d/d_0) \tag{5.2}$$

where the intercept PL_0 is the path loss at an arbitrary reference distance d_0. Such models are referred to as empirical one-slope models and are countless in the literature. For instance, the above Friis equation leads to:

$$PL(\text{dB}) = 32.44 + 20 \times \log(f/f_0) + 20 \times \log(d/d_0) \tag{5.3}$$

where $f_0 = 1\,\text{MHz}$ and $d_0 = 1\,\text{km}$.

One such model by Okumura [7] was derived from extensive measurements in urban and suburban areas. It was later put into equations by Hata [8]. This Okumura–Hata model, valid for 150 MHz to 1.5 GHz, was later extended to PCS frequencies, 1.5–2 GHz, by the COST project [9,10] and is referred to as the COST 231-Hata model; it is still widely used by cellular operators. The model provides good path loss estimates for large urban cells (1–20 km) and a wide range of parameters like frequency, base station height (30–200 m), and environment (rural, suburban, or dense urban).

Another popular model is that of Walfish–Ikegami [11,12], which was also revised by the COST project [9,10] into a COST 231-Walfish–Ikegami model. It is based on considerations of reflection and scattering above and between buildings in urban environments. It considers both line-of-sight (LOS) and nonline-of-sight (NLOS) situations. It is designed for 800 MHz to 2 GHz, base station heights of 4–50 m, and cell sizes up to 5 km, and is especially convenient for predictions in urban corridors.

More recently, Erceg [13] proposed a model derived from a vast amount of data at 1.9 GHz, which makes it a preferred model for PCS and higher frequencies [14]. These models and their applications and domains of validity are well described and analyzed, for instance, in Refs. 15–18. They provide a first estimate used by service providers in wireless systems' design phase.

Further refinements to these models in which multiple path loss exponents, $n_1, n_2, \ldots$, are used at different separation ranges provide some improvements, especially in heavy multipath indoor environments. It turns out, however, that variations from site to site are such that these multiple slope improvements are fairly small, and simple one-slope models are a good enough first approximation for outdoor propagation models. More detailed, site-specific models are required for better results, but require additional efforts and site-specific terrain or building data.

Two important points should be kept in mind about most propagation models though. The first is that large amounts of empirical data were collected at cellular and PCS frequencies (800 and 1900 MHz), and extensions to other frequencies may not have been well verified.* The second is that these data points were collected while driving and may not accurately reflect fixed wireless links, which is discussed in more detail in the following section.

5.2.2   Fixed Broadband Access

Since our focus is on fixed broadband access, we should emphasize that the propagation modeling of a fixed radio link has some fundamental differences with that of a mobile link.

The problem of collecting fixed data for an empirical model is not trivial; and many experimenters present methods to locally average data (over one-half of a wavelength) to remove small-scale fading due to multipath. Small-scale fading is difficult to quantify accurately, and even a large number of fixed data points would provide insufficient sampling to be able to evaluate its impact.

Another important issue is that of antenna beamwidth (or directivity). Mobile data collections are conducted using an omnidirectional antenna (isotropic with respect to azimuth). It has long been known that the antenna beamwidth and more specifically the distribution of angles of arrival with respect to the direction of motion of a mobile are important parameters to quantify the fading of a mobile link [16].

Consequently, fixed data models may differ in some cases from the usual empirical models. One contribution to IEEE 802.16 [14] analyzes these details and proposes models based on a large PCS data campaign and associated model [13].

Good fixed models would be welcome by the industry, but the current use of cellular and PCS models is likely to continue for a number of reasons: first, they provide a good estimate for initial design (site-specific models and simulations are used for more precise predictions); second,

---

* Typically, some frequency extensions may be obtained by adding a frequency dependence in $f^{2.6}$ (or a $26 \times \log f$ term in dB) as suggested by Ref. 19, and used for instance in the Okumura–Hata model [8] and the 802.16 contribution [14].

some time is necessary to roll out large fixed wireless systems that can be used and analyzed to provide a wide modeling range; lastly, by the time these fixed models exist, the focus of WiMAX is likely to turn again toward mobility.

5.2.3   Link Budgets

Link budgets are essential for radio systems coverage and performance predictions. Unfortunately, they depend largely on suppliers' data and are often kept proprietary. Still, important common parameters valid for most fixed WiMAX systems are given in this section. Mobile WiMAX systems require a different link budget analysis and is not covered here.*

Radio parameter values are presented here for current fixed WiMAX systems [1,2]. Some of these values, such as transmitted power and antenna gain, may change with local regulations; others, like received sensitivity, are commonly discussed in the standard community and accepted as a minimum standard that suppliers must adhere to. Of course suppliers may improve upon such numbers.

A variety of diversity schemes may be employed in WiMAX systems; they have a significant impact on link budgets. Some early systems do not use any diversity; others use simple spatial or polarization diversity schemes; and some use advanced MIMO systems.

5.2.4   Propagation Characteristics

Between transmitter and receiver, the wireless channel is modeled by several key parameters. These parameters vary significantly with the environment, rural versus urban, or flat versus mountainous. Different kinds of fading occur; they are often categorized into three types [15,16]:

Small-scale fading causes great variation within a half wavelength. It is caused by multipath and moving scatterers. Resulting fades are usually approximated by Rayleigh, Ricean, or similar fading statistics.[†] Radio systems rely on diversity, equalizing, channel coding, and interleaving schemes to mitigate its impact.

Large-scale shadowing causes variations over larger areas because of terrain, building, and foliage obstructions; its impact on link budgets is detailed further in this section.

Distance dependence is approximated by $PL = 10n \times \log(d)$, where n is the path loss exponent that varies with terrain and environment.

---

* Elements of mobile WiMAX are given, for instance, in Ref. 4, pp. 32–34.
† Analyses in many published papers also show that Nakagami-m and Weibull distributions also lead to interesting results and convenient approximations.

We will see later in Section 5.3.4 that n itself typically follows a Gaussian distribution.

The large-scale fading due to various obstacles is commonly accepted to follow a log-normal distribution [18,20,21]. This means that its attenuation x measured in dB is normally distributed $N(m, \sigma)$ with mean $m = \bar{x}$ and standard deviation σ. The probability density function of x is given by the usual Gaussian formula

$$p(x) = \frac{1}{\sigma\sqrt{2\pi}} \times \exp \frac{-(x - \bar{x})^2}{2\sigma^2} \tag{5.4}$$

With this Gaussian distribution model, the probability that the received power x at a distance d exceeds a threshold x_0 (the receiver threshold that provides an acceptable signal) is given by Ref. 22.

$$P(x \geq x_0) = \frac{1}{2}\text{erfc}\left(\frac{x_0 - \bar{x}}{\sigma\sqrt{2}}\right) \tag{5.5}$$

where erfc is the complementary error function.* Equation 5.5 is used to choose a fade margin, or excess margin, in a link budget to obtain a target service reliability (percentage of acceptable signal at the edge of planned coverage). Without that excess margin, link budgets and propagation models only yield a median propagation loss, corresponding to 50% edge coverage reliability.†

The mean of log-normal shadowing is usually incorporated in path loss model and its standard deviation σ is typically estimated by empirical measurements. Commonly accepted values for σ are between 6 and 12 dB. Measured values of σ seem to display Gaussian distribution as well and depend on: the radio frequency, the type of environment (rural, suburban, or urban), and base station and subscriber station height. Reports may be found in the literature [20–29] and are summarized in Table 5.1. The choice is somewhat arbitrary, but given the above experimental data we chose to follow an empirical value for suburban environment of $\sigma = 9.6$ dB (e.g., for terrain category B in Ref. 13) and use that same estimate $\sigma = 9.6$ dB for 3.5 GHz and 5.8 GHz. We then chose a fade margin or excess margin for a certain service reliability. For instance, service providers tend to impose a requirement of 90% edge coverage, which when following Jakes' method [22] yields a fade margin of 12.3 dB.

---

* The complementary error function is defined as $\text{erfc} = 1 - \text{erf}$, where erf is the error function $\text{erf}(x) = \frac{2}{\sqrt{\pi}} \int_0^x e^{-u^2} du$.

† Indeed, setting the excess margin to $x_0 - m = 0$ yields a coverage probability of $P(x \geq \bar{x}) = 50\%$, since $\text{erfc}(0) = 1$.

TABLE 5.1

Path Loss Exponent (n) and Log-Normal Shadowing Standard Deviation
(σ, in dB)

Source	Frequency (GHz)	Path Loss Exponent n	σ (dB)	Comments
Seidel [23]	0.9	2.8	9.6	Suburban (Stuttgart)
Erceg [13]	1.9	4.0	9.6	Terrain-category B
Feuerstein [24]	1.9	2.6	7.7	Medium antenna height
Abhayawardhana [25]	3.5	2.13	6.7–10	Ref. 25, tables 2 and 3
Durgin [26]	5.8	2.93	7.85	Ref. 26, figure 7, residential
Porter [27]	3.7	3.2	9.5	Some denser urban
Rautiainen [28]	5.3	4.0	6.1	Ref. 28, figures 3 and 4
Schwengler [29]	5.8	2.0	6.9	LOS
	5.8	3.5	9.5	NLOS
	3.5	2.7	11.7	See Section 5.3.4
Average	3.5–5.8	3.0	8.7	

Summary of values for various frequencies reported for suburban or residential areas.

TABLE 5.2

WiMAX Reverse Link Budget at 3.5 GHz, for 3.5 MHz Channels, in Different
Modulations (BPSK to 64QAM)

Parameter	Unit	Equation	BPSK 1/2	64QAM 3/4
Data rate	Mbps	r	1.4	12.7
Subscriber Tx power	dBm	A	23.0	23.0
Subscriber antenna gain	dBi	B	18.0	18.0
Subscriber cable loss	dB	C	0.0	0.0
Transmitted EIRP	dBm	$D = A + B - C$	41.0	41.0
Base Rx antenna gain	dBi	E	17.0	17.0
Base cable loss	dB	F	1.0	1.0
Thermal noise	dBm/Hz	$10 \times \log(kT) + 30$	−174.0	−174.0
Channel width	MHz	G	3.5	3.5
Thermal noise in channel	dBm	$H = 10 \times \log(kTG) + 90$	−108.6	−108.6
Base noise figure	dB	I	4.0	4.0
Base noise floor	dBm/Hz	$J = H + I$	−104.6	−104.6
SNR required	dB	K	6.4	24.4
Receiver interference margin	dB	L	0.0	0.0
Base Rx sensitivity	dBm	$M = J + K + L$	−98.2	−80.2
Diversity gain	dB	N	0.0	0.0
Total System gain	dB	$Q = D + E - F - M + N$	155.2	137.2
Log-normal fading std dev	dB	σ	9.6	9.6
Log-normal fade margin	dB	O	12.3	12.3
Building penetration loss	dB	P	0.0	0.0
Maximum reverse path loss	dB	$R = D + E - F - M + N - O - P$	142.9	124.9

TABLE 5.3

WiMAX Reverse Link Budget at 5.8 GHz, for 10 MHz Channels, in Different Modulations (BPSK to 64QAM)

Parameter	Unit	Equation	BPSK 1/2	64QAM 3/4
Data rate	Mbps	r	2.0	18.2
Subscriber Tx power	dBm	A	18.0	18.0
Subscriber antenna gain	dBi	B	16.0	16.0
Subscriber cable loss	dB	C	0.0	0.0
Transmitted EIRP	dBm	$D = A + B - C$	34.0	34.0
Base Rx antenna gain	dBi	E	16.0	16.0
Base cable loss	dB	F	1.0	1.0
Thermal noise	dBm/Hz	$10 \times \log(kT) + 30$	−174.0	−174.0
Channel width	MHz	G	10.0	10.0
Thermal noise in channel	dBm	$H = 10 \times \log(kTG) + 90$	−104.0	−104.0
Base noise figure	dB	I	4.0	4.0
Base noise floor	dBm/Hz	$J = H + I$	−100.0	−100.0
SNR required	dB	K	6.4	24.4
Receiver interference margin	dB	L	0.0	0.0
Base Rx sensitivity	dBm	$M = J + K + L$	−93.6	−75.6
Diversity gain	dB	N	0.0	0.0
Total system gain	dB	$Q = D + E - F - M + N$	142.6	124.6
Log-normal fading std dev	dB	σ	9.6	9.6
Log-normal fade margin	dB	O	12.3	12.3
Building penetration loss	dB	P	0.0	0.0
Maximum reverse path loss	dB	$R = D + E - F - M + N - O - P$	130.3	112.3

We summarize parameters for licensed radio systems at 3.5 GHz with the link budget shown in Table 5.2.

Link budgets in unlicensed bands are similar to the above but are usually limited by a lower maximum allowed EIRP as shown in Table 5.3.

5.2.5 In-Building Penetration

Fixed wireless service may use antennas placed on individual homes, but that comes with a number of obvious problems: customers may not welcome structures on their homes, and installation time and cost are high. The holy grail of wireless access consists in shipping a small device, like ADSL or cable modem, that customers may install without on-site technician time. Furthermore, the clear advantage of wireless data services lies in its portability or full mobility; therefore it seems clear that the trend is to pursue small indoor devices. Unfortunately, sending RF signal into buildings comes at an additional cost that can be quantified by an additional building penetration loss in the link budget.

Measurement campaigns show once again that the distribution is close to log-normal [20]. A Gaussian function is a good approximation of the cumulative distribution function (CDF) of indoor measurements, as plotted in

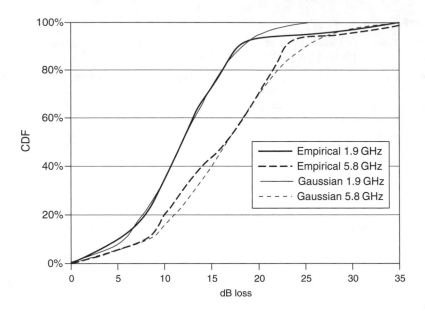

FIGURE 5.1
Penetration loss into residential buildings, cumulative density distribution, and Gaussian
approximation for 1900 and 5800 MHz.

Figure 5.1. The mean and standard deviations of indoor penetration loss vary
with frequency, types of homes, and environment around the homes. Variations also depend on the location within the building (near an outside wall,
a window, or further inside). Finally, the angle of incidence with the outside
wall also has a significant impact [30]. Precise characterization of in-building
penetration is therefore difficult. Nonetheless, an approximation of an average penetration loss μ_i around 12–15 dB and a standard deviation σ_i between
5 and 8 dB seems to be the norm in published studies [26,31,32]. Table 5.4
summarizes some published results for residential homes.

Many similar studies are available for university or industrial campuses as
well as high-rises, but these values are typically higher than for residential
homes. They also depend heavily on the floor, height of neighboring buildings, or clutter. Let us limit our analysis to residential and suburban areas.
Few measurements are available at 3.5 GHz. The review of fairly large data
collection campaigns at 1.9, 2.5, and 5.8 GHz [29–33], as well as personal
measurements are summarized in Figure 5.1 and in Table 5.4. These results
lead us to choose empirical values of $\mu_i = 12$ dB at 3.5 GHz, $\mu_i = 15$ dB at
5.8 GHz, and $\sigma_i = 6$ dB in both cases.

With that in mind, we consider that in-building penetration is a log-normal
random variate independent of the large-scale shadowing. Therefore, the
log-normal fading used for indoor propagation should be the normal random
variable $N(\mu_i, \sqrt{\sigma^2 + \sigma_i^2})$. Both median penetration loss and modified excess
margin should be taken into account for a new indoor link budget.

TABLE 5.4

Penetration Loss into Residential Buildings: Median Loss (μ_i) and Standard Deviation (σ_i) from Experimental Results Reported at Various Frequencies

Source	Frequency (GHz)	μ_i (dB)	σ_i (dB)	Comments
Aguirre [31]	1.9	11.6	7.0	Ref. 31, figure 3
	5.9	16.1	9.0	
Durgin [26]	5.8	14.9	5.6	Ref. 26, table 5 average
Martijn [32]	1.8	12.0	4.0	Ref. 32, table 1
Oestges [30]	2.5	12.3	–	Ref. 30, table 6 (avg. $L_e + L'_{ge}$)
Schwengler	1.9	12.0	6.0	Personal measurements
Schwengler [29]	5.8	14.7	5.5	Ref. 29, table 2
Average	≈ 2	12.0	5.7	
	5.8	15.2	6.7	

TABLE 5.5

WiMAX Reverse Link Budget at 3.5 GHz into Residential Buildings, for 3.5 MHz Channels, in Different Modulations (BPSK to 64QAM)

Parameter	Unit	Equation	BPSK 1/2	64QAM 3/4
Data rate	Mbps	r	1.4	12.7
Subscriber Tx power	dBm	A	23.0	23.0
Subscriber antenna gain	dBi	B	18.0	18.0
Subscriber cable loss	dB	C	0.0	0.0
Transmitted EIRP	dBm	$D = A + B - C$	41.0	41.0
Base Rx antenna gain	dBi	E	17.0	17.0
Base cable loss	dB	F	1.0	1.0
Thermal noise	dBm/Hz	$10 \times \log(kT) + 30$	−174.0	−174.0
Channel width	MHz	G	3.5	3.5
Thermal noise in channel	dBm	$H = 10 \times \log(kTG) + 90$	−108.6	−108.6
Base noise figure	dB	I	4.0	4.0
Base noise floor	dBm/Hz	$J = H + I$	−104.6	−104.6
SNR required	dB	K	6.4	24.4
Receiver interference margin	dB	L	0.0	0.0
Base Rx sensitivity	dBm	$M = J + K + L$	−98.2	−80.2
Diversity gain	dB	N	12.0	12.0
Total system gain	dB	$Q = D + E - F - M + N$	167.2	149.2
Combined log-normal std dev	dB	$\sqrt{\sigma^2 + \sigma_i^2}$	11.3	11.3
Log-normal Fade Margin	dB	O	14.4	14.4
Building Penetration Loss	dB	P	12.0	12.0
Maximum Reverse Path Loss	dB	$Q = D + E - F - M + N - O - P$	140.8	122.8

This has a significant impact on the total link budget—see Table 5.5. In fact, some manufacturers even claim that indoor devices are impractical in unlicensed bands, which would lead to too small a radii of coverage in the limited unlicensed power levels. In licensed bands as well, even though higher transmit power is allowed, indoor radio units need to somehow increase their link budgets: advanced diversity schemes with a plurality of antennas are usually used. Some WiMAX systems also have the ability to use sub-channel groups with a dynamic number of subcarriers; link budget may then be increased by providing full power to that group (at the cost of overall throughput).

That same argument may be made for unlicensed frequencies as well; advanced diversity combining schemes and MIMO may be enough to compensate for high penetration losses as well as for the low transmit powers allowed [34].

5.3   System Performance

Service providers are in an intensive phase of trials and performance evaluations for fixed WiMAX systems and services. Initial technical evaluation showed promising data rates and a number of more wide-scale trials were conducted on a larger customer base throughout the world—in Europe, Asia, and the Americas.

5.3.1   Data Rates

IEEE 802.16 and WiMAX profiles allow for several radio channel bandwidths, which lead to very different data rates. In a given profile, physical layer data rate of a WiMAX system is determined by the type of modulation and coding: from BPSK 1/2 to QAM64 3/4. Theoretical data rates are quoted in standards or by manufacturers but actual throughput vary with suppliers: a degradation of 40%–50% is often observed. Table 5.6 summarizes typical data rates observed in a 3.5 MHz FDD channel (also see Figure 5.6). That seemingly large difference is mainly due to timing delays necessary for scheduling and collision avoidance between users. Actual data results vary with suppliers, and interoperability between suppliers introduce even greater variations. Nevertheless, the great value of WiMAX-certified products is to guarantee some minimum performance: a service provider may rely on the fact that WiMAX-certified products will work well with other suppliers certified for the same profile.

These results are for one direction 3.5 MHz channel, a full duplex FDD system may see up to twice as much throughput in the total 7 MHz bandwidth. Of course, different profiles and channel widths lead to different throughput results. An unlicensed TDD 10 MHz channel profile for instance has the advantage of adapting to asymmetrical data demand. Similar benchmark tests show that such a system is also capable of throughputs of around 8 Mbps (see Figure 5.7).

TABLE 5.6

WiMAX 3.5 MHz Channel Maximum Theoretical and Actual Measured Throughput (at 3.5 GHz)

Modulation	3.5 MHz sensitivity (dBm)	SNR (dB)	Theoretical (Mbps)	Actual (Mbps)
BPSK 1/2	−90.6	6.4	1.41	0.86
BPSK 3/4	−88.6	8.5	2.1	1.28
QPSK 1/2	−87.6	9.4	2.82	1.72
QPSK 3/4	−85.8	11.2	4.23	2.58
16QAM 1/2	−80.6	16.4	5.64	3.44
16QAM 3/4	−78.8	18.2	8.47	5.16
64QAM 2/3	−74.3	22.7	11.29	6.88
64QAM 3/4	−72.6	24.4	12.71	7.74

TABLE 5.7

Typical Parameters for SUI-1 to 6 Channel Models

Channel Model	Terrain Type	RMS Delay Spread (μs)	Doppler Shift	Ricean K factor (dB)
SUI-1	C	0.042 (Low)	Low	14.0
SUI-2	C	0.069 (Low)	Low	6.9
SUI-3	B	0.123 (Low)	Low	2.2
SUI-4	B	0.563 (High)	High	1.0
SUI-5	A	1.276 (High)	Low	0.4
SUI-6	A	2.370 (High)	High	0.4

Delay spread values estimated for 30-degree antennas azimuthal beamwidths, and ricean K-factors are for 90% cell coverage.
Source: IEEE 802.16 Broadband Wireless Access Working Group, 2003.

Interferences from other cells (cochannel interferences) strongly impact actual rates [35,36]. And in unlicensed cases, unwanted interferences in the band are also a concern: minimum signal to noise ratios listed in Table 5.6 must be maintained for a given throughput.

To compare system performance in diverse environments, tests are usually conducted with traffic load generators and fading emulators. Service providers can thus create repeatable benchmark tests, in a controlled environment, to compare equipment performance under different conditions. These tests quantify the different access performances in large rural areas, suburban areas, or dense urban cores, both for fixed access and full mobility.

Stanford University Interim (SUI) models are used to create a small number of models that emulate different terrain types, Doppler shift, and delay spread as summarized in Table 5.7. Terrain types are (from Ref. 13) defined as follows: the maximum path loss category (A), hilly terrain with moderate-to-heavy tree densities; the intermediate path loss category (B), hilly with light tree density or flat with moderate-to-heavy tree density; the minimum path loss category (C), mostly flat terrain with light tree densities. In some cases, these terrain categories are used to refer to obstructed urban, low-density suburban, and rural environments, respectively.

5.3.2   Experimental Data

As an example, let us illustrate the above with data for fixed broadband access in a residential suburban area. Unlike mobile cellular systems, a fixed wireless access system needs a careful selection process for qualifying customers. Propagation tools and terrain data are used in that process, but the level of detail is a matter of choice. A precise qualification process leads to better targeted mailing and may avoid miscalculated predictions. Service providers cannot afford to be too optimistic nor too pessimistic in their predictions: false negatives are a missed revenue opportunity, and false positives lead to wasted technician time and unhappy customers. It is therefore time well spent to refine selection criteria and tools as much as possible.

A simple selection process consists of geocoding customers' addresses and correlating them to terrain data as well as to a simple propagation model for an initial estimate. Address geocoding, however, is far from a perfect process. A customer address may not give reliable longitude and latitude, and will rarely hint on where an outdoor antenna may be in good RF visibility of a base station. Some manual processing and even some local knowledge of the area may be required; and in the end, a site visit may still discard a possible location. The quality of terrain data and RF modeling is of course also of high importance. Terrain data can be obtained at no cost from U.S. geological surveys (100 or 30 m accuracy), which is useful for path loss prediction, but it will not accurately predict shadowing in all areas. More granular data, including building data, with submeter accuracy can be obtained at a much higher cost. Another alternative is to drive-test around the area of interest and to optimize a propagation model in a given area. Many software packages allow for such model optimization, which significantly improve prediction tools. (Of course these models, as well as the drive-test optimizations, are usually based on mobile data.)

5.3.3   Other Trial Considerations

In many cases, trial data are published and compared to existing models or (if extensive enough) used to create a new propagation model. Many other aspects of major customer trials are important to service providers, such as: customer qualification, installation, support, troubleshooting, and overall estimation of customer satisfaction.

- The overall trial goal makes a significant difference in trial results: the customer selection process for instance may focus on capacity limitations in a specific area, or it may be geared toward testing distance limits of a radio system; clearly trial results will be different.
- Trial architectures vary. Most WiMAX radio systems use Ethernet network interfaces, but many systems require a mixture of backhaul

or longhaul transport, which include microwave, copper, or fiber links, over TDM T1, T3, SONET, etc.

- Integration to a monitoring system is also a major portion of a technical trial. Major network element (including customer devices) should be monitored. Maintenance, repairs, and upgrades should be performed in a low-intrusion maintenance window to limit the impact of downtime.

- Most network elements should be controlled remotely and centrally from a network operations center. Good control of network elements, including customer equipment, is precious for system support, especially when it reaches large scale.

- Data collection is highly important for a trial. As a successful trial moves into production, ongoing data analyses are still important for network optimization.

- Customer satisfaction surveys and focus groups are also an integral part of a complete trial; they should also continue into production phase and be compared to network quality metrics.

5.3.4 Radio Parameters Analysis and Modeling

In an initial design phase, a simple one-slope model and low-resolution terrain data suffice for a rough estimate to qualify customers. As operations progress, actual measurements should be compared to predictions and the process is refined.

For instance, an initial selection process leads to the chart on Figure 5.2. Actual measurements show the right trend, but some variations are very large (sometimes in excess of 20 dB). Better modeling and drive testing should be considered in this case.*

During trials, a received signal strength indicator (RSSI), in dBm, is logged at all customer locations. A plot of RSSI as a function of the logarithm of distance is graphed in Figure 5.3. The logarithmic scale for the distance is simply justified by the fact that a one-slope model will show a linear approximation on the graph. Many propagation studies use this scale since it allows for easy comparison of path loss exponents. The variations in RSSI for a given customer location are represented by error bars at each point. Each error bar represents a standard deviation; that is, the total width of the error bar shows two standard deviations.

The next step in data analysis is a comparison between the data set and typical models. For that comparison, a path loss estimate should be derived from the empirical system. The RSSI measurement provides one term of the path loss. The other is in the transmitted power level, which depends on base

---

* The linear approximation of scatter plot in Figure 5.2 does not cross the axis at zero; the line is offset by almost 5 dB due to some fixed system differences between actual and measured values. The slope of the line is 1 as it should be.

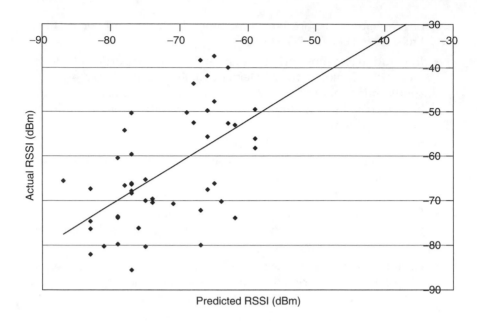

FIGURE 5.2
Actual RSSI measured at customer locations versus predicted RSSI from the planning model.

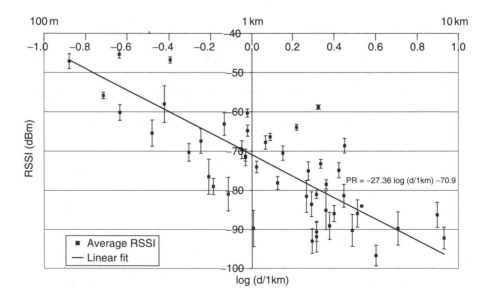

FIGURE 5.3
Received power level signal strength indicator (in dBm) as a function of distance (on a logarithmic scale).

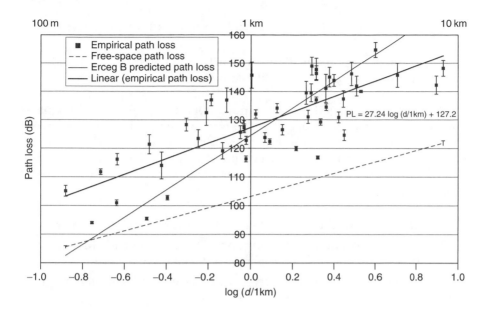

FIGURE 5.4
Empirical path loss as a function of distance (on a logarithmic scale) and comparison to
prediction models.

station power, cable loss, antenna pattern, and even (to a small extent) on
the deviation from boresight of the sector's antenna.* Path loss estimates are
represented in Figure 5.4.

Approximation of path loss to a one-slope model leads to the following
equation:

$$PL(\text{dB}) = 127.2 + 27.24 \times \log(d/d_0) \qquad (5.6)$$

with $d_0 = 1$ km. The trial environment is compared to typical cellular models
as discussed below.

- Path loss exponent is approximately $n = 2.7$. The Walfish–Ikegami
  model for line-of-sight in urban corridors predicts $n = 2.6$. Other
  reports have shown similar results for 3.5 GHz: Ref. 25 reports values of n between 2.13 and 2.7 for rural and suburban environments,
  Ref. 27 reports $n = 3.2$. However, many other models predict higher

---

* Deviation from boresight may be easily estimated for fixed access where customer locations
were previously geocoded. From geocoded data, a bearing with respect to the serving base
combined with the known orientation of the sectors antennas yield an angle off boresight for every
customer. A specific attenuation number can then be included for a better path loss estimate. In
most designs, sectors will overlap around the 3 dB beam width, and omitting this term would not
result in more than 3 dB error in the path loss estimate. Nevertheless, the calculations involved
are easy enough to improve the path loss estimate.

exponents for n, between 3.5 and 4.5 (see path loss exponents in Table 5.1).

- Otherwise, approximations are fairly good with Erceg-B and C models. Erceg-B is the best fit and is represented in Figure 5.4.

The most popular method to compute slope estimate is a least squares error estimate. In that method, a set of error terms $\{e_i\}$ is defined between each data point and a linear estimate. Minimizing the sum of these errors yields the slope and intercept, which intuitively gives a good approximation of the data set. That method also benefits from the following important properties [37]:

1. Least squares estimated slope and intercept are unbiased estimators.
2. Standard deviations of the slope and intercept depend only on the known data points and the standard deviation of the error set $\{e_i\}$.
3. Estimated slope and intercept are linear combinations of the errors $\{e_i\}$.

From the last point, if we assume that the errors are independent normal random variables (as in a log-normal shadowing situation), the estimated slope and intercept are also normally distributed. If we assume more generally that the data points are independent, the central limit theorem implies that for large data sets, the estimated slope and intercept tend to be normally distributed.

For the last assumption to be true, very low correlation of the wireless channel must exist between data points. This is the case when data points are measured at fixed locations tens or hundreds of meters apart—in which case measurements show very low correlations between the respective fading channels. Similarly, this is the case even in a mobile cellular environment, from one cell to another.

The important conclusion is that path loss exponent is approximated by a normal (or Gaussian) random variable.

We also verify a few more key findings as in Ref. 13, for a 3.5 GHz fixed link:

1. Free-space approximation $(PL_0 = 20 \times \log(4\pi d_0/\lambda))$ works well within 100 m.
2. Path loss exponent depends strongly on height of transmitter (mobile height being more or less constant throughout).
3. Variations around median path loss are Gaussian within a cell (log-normal shadowing) with standard deviation $\sigma \approx 11.7$ dB.
4. Unfortunately, our limited number of cells do not allow us to quantify the nature of the variations of σ over the population of macro cells.

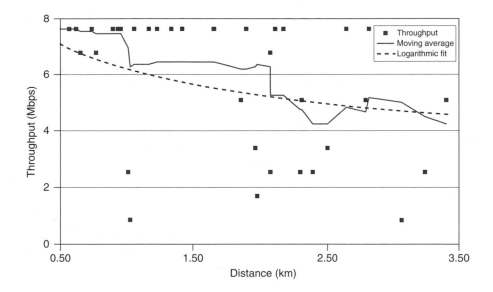

FIGURE 5.5
Throughput measured at customer locations as a function of distance to base station with ten point moving average and logarithmic fit.

5.3.5 Throughput Measurements

Having now characterized RF levels, we focus on the parameter of most interest: data throughput. Throughput is affected by distance, shadowing, and interferences. The parameter of importance is the signal to noise ratio (SNR); it can be estimated from RSSI and ambient noise measurements or can usually be reported in some form by the RF equipment. The SNR has a direct impact on the modulation used by the link* and therefore on the throughput of that link. That throughput is graphed as a function of distance in Figure 5.5.

In fact, modulation and throughput change from time to time. It may be important to study the statistical distribution of the resulting throughput, as in Figures 5.6 and 5.7. These figures show the probability of reaching a certain throughput, over the population of fixed location under test. These plots may be compared to plots representing fixed modulations and controlled fading environments described in Section 5.3.1. Fading statistics in suburban areas shows close correlation with SUI models 3 and 5, and throughput density functions near those of 16QAM 3/4 in such fading environments [38].

Finally, we report on the standard deviation of measured signal strength. In most cellular trials mobile data is collected, which makes it impossible to quantify variations over long periods of time for a given location. In a population of fixed location, however, a measured standard deviation over a long

---

* The details of that correlation are far from simple and depend greatly on the suppliers' implementation choices.

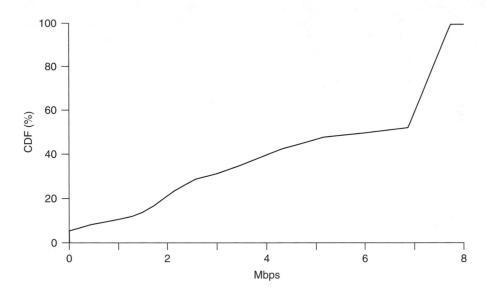

FIGURE 5.6
Throughput cumulative distribution statistic measured in a 3.5 MHz FDD channel at 3.5 GHz.

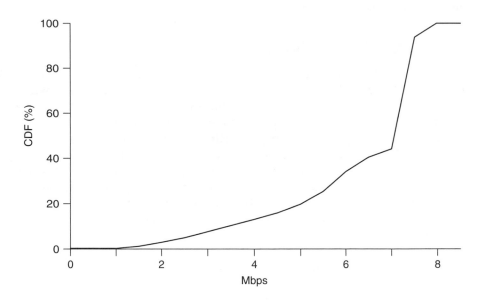

FIGURE 5.7
Throughput cumulative distribution statistic measured in a 10 MHz TDD channel at 5.8 GHz.

period may be useful in predicting seasonal changes in the radio channel. Typical standard deviations in fixed links over several months vary between 1 and 6 dB; when deciduous trees are present, the value increases in the spring as leaves come out. Trial data also show that the standard deviation tends to

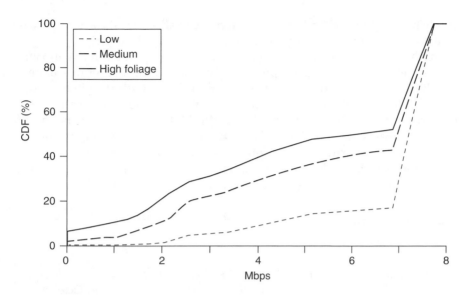

FIGURE 5.8
Throughput cumulative distribution statistic measured in various foliage conditions for a fixed links in a given area, for a 3.5 MHz FDD channel at 3.5 GHz.

increase with distance. A median value of the standard deviation of path loss is given by

$$\sigma_{\text{fixed}} = 2.26 + 0.75 \times \log(d/d_0) \tag{5.7}$$

with $d_0 = 1$ km.

Seasonal variations are especially noticeable as leaves come out. The impact on the link budget has been reported for fixed wireless links [39] and in different wind conditions [40]. We measure some variations of the path loss exponent, the intercept, and the log-normal shadowing. In many cases, the wireless system can adapt to these variations, but in some marginal locations where link budget nears the maximum allowable path loss, throughput is affected. As shown on Figure 5.8, low bit rates are affected the most by changes in foliage.

5.4 Conclusion

Modern wireless communications improve continuously in performance and availability, but still require good design methods based on the fundamentals of radio propagation. We reviewed important aspects of propagation modeling for mobile and fixed wireless access; we quantified performance of fixed WiMAX systems in residential surroundings; and we compared them to other published trial results and models. Extensive trial results show that WiMAX offers good opportunities for broadband wireless applications.

Analysts and strategists have been announcing ubiquitous broadband wireless services for years now, yet pessimists claim that these services will never see the light of day. Still, eventually a combination of events will be the catalyst for the broadband wireless industry: new technology advances, new spectrum bands, efficient standards like WiMAX, good conformance certification processes, flexible IP-based network infrastructure, involvement from major chip manufacturers, and global economies of scale are all encouraging signs. One can hope that wireless service providers will deploy these new services in most cities and even in lower-density suburban and rural areas.

References

1. IEEE Std 802.16-2004, *IEEE Standard for Local and Metropolitan Area Networks—Part 16: Air Interface for Fixed Broadband Wireless Access Systems*, October 2004. Available at http://standards.ieee.org/getieee802/802.16.html.
2. IEEE Std 802.16/Conformance03-2004, *IEEE Standard for Conformance to IEEE Std 802.16—Part 3: Radio Conformance Tests (RCT) for 10–66 GHz WirelessMAN-SC Air Interface*, June 2004. Available at http://standards.ieee.org/getieee802/802.16.html.
3. IEEE Std 802.16e-2005, *IEEE Standard for Local and Metropolitan Area Networks—Part 16: Air Interface for Fixed and Mobile Broadband Wireless Access Systems*, February 2006. Available at http://standards.ieee.org/getieee802/802.16.html.
4. WiMAX Forum (August 2006), *Mobile WiMAX—Part I: A Technical Overview and Performance Evaluation*. Available at www.wimaxforum.org.
5. WiMAX Forum (May 2006), *Mobile WiMAX—Part II: A Comparative Analysis*. Available at www.wimaxforum.org.
6. P. Papazian, M. Cotton, *Relative Propagation Impairments between 430 MHz and 5750 MHz for Mobile Communication Systems in Urban Environments*, NTIA Report TR-04-407, December 2003.
7. Y. Okumura, E. Ohmori, T. Kawano, K. Fukuda, Field strength and its variability in VHF and UHF land-mobile radio service, *Review of the Electrical Communication Laboratory*, Volume 16, Nos. 9–10, pp. 825–873, September–October 1968.
8. M. Hata, Empirical formula for propagation loss in land mobile radio services, *IEEE Transactions on Vehicular Technology*, Volume 29, No. 3, pp. 317–325, August 1980.
9. European Cooperation in the Field of Scientific and Technical Research, EURO-COST 231, *Urban Transmission Loss Models for Mobile Radio in the 900 and 1800 MHz Bands*, COST 231 TD (91) 73. Rev 2, The Hague, September 1991.
10. European Cooperation in the Field of Scientific and Technical Research, EURO-COST 231, *Digital Mobile Radio Towards Future Generation Systems*, COST 231 Final report. Available at http://www.lx.it.pt/cost231/.
11. F. Ikegami, S. Yoshida, T. Takeuchi, M. Umehira, Propagation factors controlling mean field strength on urban streets, *IEEE Transactions on Antennas & Propagation*, Volume AP32, pp. 822–829, 1984.

12. J. Walfish, H.L. Bertoni, A Theoretical model of UHF propagation in urban environment, *IEEE Transactions on Antennas & Propagation*, Volume AP-36, pp. 1788–1796, December 1988.
13. V. Erceg, L.J. Greenstein, S.Y. Tjandra, S.R. Parkoff, A. Gupta, B. Kulic, A.A. Julius, R. Bianchi, An empirically based path loss model for wireless channels in suburban environments, in *IEEE Journal on Selected Areas in Communications*, Volume 17, No. 7, July 1999.
14. IEEE 802.16 Broadband Wireless Access Working Group, *Channel Models for Fixed Wireless Applications*, contribution to 802.16a, 2003. Available at http://wirelessman.org/tga/docs/80216a-03_01.pdf.
15. A. Goldsmith, *Wireless Communications*, Cambridge University Press, New York, 2005.
16. T.S. Rappaport, *Wireless Communications: Principles and Practice—Second Edition*, Prentice-Hall, New Jersey, 2002.
17. W.C.Y. Lee, *Wireless and Cellular Communications, 3rd ed.*, McGraw Hill, New York, 2005.
18. H.L. Bertoni, *Radio Propagation for Modern Wireless Systems*, Prentice-Hall Inc., New Jersey, 2000.
19. T.-S. Chu, L.J. Greenstein, A quantification of link budget differences between the cellular and PCS bands, *IEEE Transactions on Vehicular Technology*, Volume 48, No. 1, pp. 60–65, January 1999.
20. C. Chrysanthou, H.L. Bertoni, Variability of sector averaged signals for UHF propagation in cities, *IEEE Transactions on Vehicular Technology*, Volume 39, Issue 4, pp. 352–358, November 1990.
21. L.J. Greenstein, V. Erceg, Y.S. Yeh, M.V. Clark, A new path-gain/delay-spread propagation model for digital cellular channels, *IEEE Transactions on Vehicular Technology*, Volume 46, Issue 2, pp. 477–485, May 1997.
22. W. Jakes, *Microwave Mobile Communications*. New York, IEEE, 1974; Reedited IEEE Press, Piscataway, 1993.
23. S.Y. Seidel, Path loss, scattering and multipath delay statistics in four european cities for digital cellular and microcellular radiotelephone, *IEEE Transactions on Vehicular Technology*, Volume 40, Issue 4, pp. 721–730, November 1991.
24. M.J. Feuerstein, K.L. Blackard, T.S. Rappaport, S.Y. Seidel, H.H. Xia, Path loss, delay spread, and outage models as functions of antenna height for microcellular system, *IEEE Transactions on Vehicular Technology*, Volume 43, No. 3, pp. 487–498, August 1994.
25. V.S. Abhayawardhana, I.J. Wassell, D. Crosby, M.P. Sellars, M.G. Brown, Comparison of empirical propagation path loss models for fixed wireless access systems, *Vehicular Technology Conference, Spring 2005*, Volume 1, pp. 73–77, 30 May–1 June 2005.
26. G.D. Durgin, T.S. Rappaport, H. Xu, Measurements and models for radio path loss in and around homes and trees at 5.85 GHz, *IEEE Transactions on Communications*, Volume 46, No. 11, pp. 1484–1496, November 1998.
27. J.W. Porter, I. Lisica, G. Buchwald, Wideband mobile propagation measurements at 3.7 GHz in an urban environment, *IEEE Antennas and Propagation Society International Symposium*, Volume 4, pp. 3645–3648, 20–25 June 2004.
28. T. Rautiainen, K. Kalliola, J. Juntunen, Wideband radio propagation characteristics at 5.3 GHz in suburban environments, in *Proc. IEEE 16th International Symposium on Personal, Indoor and Mobile Radio Communications, PIMRC 2005*, Volume 2, pp. 868–872, 11–14 September, 2005.

29. T. Schwengler, M. Gilbert, Propagation models at 5.8 GHz—path loss and building penetration, in *Proc. 2000 IEEE Radio and Wireless Conference*, pp. 119–124, 10–13 September, 2000.

30. C. Oestges, A.J. Paulraj, Propagation into buildings for broadband wireless access, *IEEE Transactions on Vehicular Technology*, Volume 53, Issue 2, pp. 521–526, March 2004.

31. S. Aguirre, L.H. Loew, L. Yeh, Radio propagation into buildings at 912, 1920, and 5990 MHz using microcells, in *Proc. 3rd IEEE ICUPC*, pp. 129–134, October 1994.

32. E.F.T. Martijn, M.H.A.J. Herben, Characterization of radio wave propagation into buildings at 1800 MHz, *Antennas and Wireless Propagation Letters*, Volume 2, Issue 1, pp. 122–125, 2003.

33. L.H. Loew, Y. Lo, M.G. Laflin, E.E. Pol, *Building Penetration Measurements From Low-Height Base Stations at 912, 1920, and 5990 MHz*, NTIA Report 95-325, September 1995.

34. D. Tse, P. Viswanath, *Fundamentals of Wireless Communications*, Cambridge University Press, New York, 2005.

35. C.F. Ball, E. Humburg, K. Ivanov, F. Treml, Performance analysis of IEEE 802.16-based cellular MAN with OFDM-256 in mobile scenarios, in *Proc. 2005 IEEE 61st Vehicular Technology Conference, VTC 2005-Spring*, Volume 3, pp. 2061–2066, 30 May–1 June, 2005.

36. F. Wang, A. Ghosh, R. Love, K. Stewart, R. Ratasuk, R. Bachu, Y. Sun, Q. Zhao, IEEE 802.16e system performance: Analysis and simulations, in *Proc. IEEE 16th International Symposium on Personal, Indoor and Mobile Radio Communications, PIMRC 2005*, Volume 2, pp. 900–904, 11–14 September, 2005.

37. J.A. Rice, *Mathematical Statistics and Data Analysis, 2nd ed.*, Duxbury Press, Pacific Grove, California 1995.

38. T. Schwengler, N. Pendharkar, Testing of fixed broadband wireless systems at 5.8 GHz, in *Proc. Technical, Professional and Student Development Workshop, 2005 IEEE Region 5 and IEEE Denver Section*, pp. 32–38, April 2005.

39. M.J. Gans, N. Amitay, Y.S. Yeh, T.C. Damen, R.A. Valenzuela, C. Cheon, J. Lee, Propagation measurements for fixed wireless loops (FWL) in a suburban region with foliage and terrain blockages, *IEEE Transactions on Wireless Communications*, Volume 1, Issue 2, pp. 302–310, April 2002.

40. M.H. Hashim, S. Stavrou, Measurements and modelling of wind influence on radiowave propagation through vegetation, *IEEE Transactions on Wireless Communications*, Volume 5, Issue 5, pp. 1055–1064, May 2006.

6

Mobility Support for IEEE 802.16e System

Hyun-Ho Choi and Dong-Ho Cho

CONTENTS

6.1   Overview of Mobility-Supporting Functions

The IEEE 802.16e system called *Mobile WiMAX* [1] has been standardized to add user mobility to the original IEEE 802.16 system (WiMAX) [2]. Since mobility causes a number of problems and requirements in wireless systems, to support user mobility, a mobile station (MS) and a base station (BS) in the mobile WiMAX system need to introduce several mobility-supporting functions to the existing WiMAX system [3].

First of all, terminals in mobile environments must rely on portable power sources, such as batteries. Since batteries provide a limited amount of energy, it is important for mobile terminals to have an efficient power-saving mechanism. The basic approach to power saving in wireless systems is discontinuous reception in which an MS periodically powers off its reception units (enters sleep state) to save power instead of continuously listening to radio channels [4]. On this basis, the IEEE 802.16e system also provides a similar sleep mode operation that provides efficient power-saving mechanisms that take into account the traffic attributes of various application services.

Second, an MS may move out of the coverage range of the current BS due to its mobility. Hence, to maintain a seamless service connection, the MS should find another BS that can serve it and establish a connection with that BS. We call this operation of transferring an ongoing connection to another BS to prevent loss or interruption of service as handover. The HO function enables the MS to have unlimited mobility and continuity of service, and hence is one of the most important functions in wireless cellular networks. The IEEE 802.16e system provides not only a basic HO function to support MS mobility, but also various techniques that enhance HO performance [5].

Finally, in cellular networks, the location of MSs is managed by two processes: paging and location update [6]. Paging is a process by which a network searches for dormant MSs by broadcasting/multicasting a paging message in predetermined areas. Location update enables MSs to inform the network of their location. A wireless system that supports the paging scheme allows MSs to operate in two modes: active mode and idle mode. If there is no traffic to or from an MS for a given period, the MS is allowed to change its mode to idle. In idle mode, the MS does not have to maintain the connection with the network and performs location update less frequently, since there is no need for the location of the MS to be traced precisely. Therefore, the MS can reduce its consumption of battery power and radio resources significantly, and the BS can eliminate unnecessary air interface and HO traffic. To allow networks to take advantage of the benefits of paging and location update, the IEEE 802.16e system also provides MSs' idle mode operation as an optional support function.

We now explain in detail the operation of the main mobility functions defined in the medium access control (MAC) layer of the IEEE 802.16e system: power-saving mechanism, HO operation, and paging and location update.

6.2 Power-Saving Mechanism

The power-saving mechanism of IEEE 802.16e enables MSs to operate in one of the two operational modes: wake mode and sleep mode [7]. In wake mode, MSs are always powered up to communicate with their serving BS, but in sleep mode they can power down to conserve energy during prenegotiated intervals. In sleep mode, there are two operational windows (i.e., time intervals): sleep window and listening window. MSs in sleep mode basically switch between the two windows. During a sleep window, they turn off most of their circuits to minimize energy consumption and so cannot receive or transmit any data. During a listening window, they synchronize with their serving BS and receive small amounts of data or a traffic indication message.

The IEEE 802.16e standard provides three kinds of power-saving class (PSC), which operate according to the characteristics of the traffic for various types of service. Each PSC uses a different operational mechanism and parameter set appropriate to the traffic characteristics. If an MS has multiple concurrent service connections, each with different traffic characteristics, it can utilize several PSCs at the same time, each of which is appropriate for a different service connection. Figure 6.1 depicts an example of sleep mode operation with two PSCs. Class A contains several connections of best effort (BE) and nonreal-time variable rate (NRT-VR) type, and Class B contains a single connection of unsolicited grant service (UGS) type. To accommodate the use by an MS of multiple PSCs, an unavailability interval is defined as a time interval that does not overlap with any listening window of any active PSC. During the unavailability interval, a BS does not transmit to the MS and buffers or drops downlink packets addressed to the MS, so the MS can power down components for physical operation. By contrast, an availability

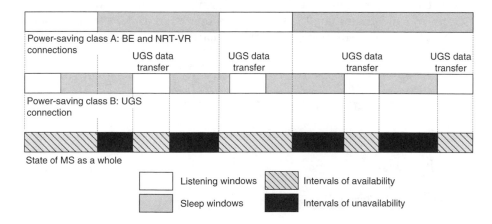

FIGURE 6.1
Example of sleep mode operation with two power-saving classes.

interval is a time interval that does not overlap with any unavailability interval. During the availability interval, the MS is expected to receive all downlink transmissions in the same way as in the wake mode.

For sleep mode operation, the IEEE 802.16e standard defines management messages as follows:

- *MOB_SLP-REQ* is transmitted from an MS to the BS and used to request the activation of PSC types I, II, and III. It contains the definition of the new PSC that is requested.

- *MOB_SLP-RSP* is sent from the BS to the MS in response to the MOB_SLP-REQ message or is sent unsolicited by the BS to activate sleep mode operation. It contains the definition of a new PSC.

- *MOB_TRF-IND* is sent from the BS to the MS by using broadcast or multicast. This message indicates whether there has been any traffic addressed to each MS that is in sleep mode. Whenever an MS enters the listening state, it wakes up, decodes this message, and confirms an indication addressed to itself.

6.2.1   Power-Saving Class of Type I

PSC of type I (PSC I) is recommended for BE and NRT-VR connections, which are used for such activities as web browsing, email, and FTP. Figure 6.2 illustrates the basic sleep mode operation of PSC I in IEEE 802.16e. To start PSC I operation, an MS sends an MOB_SLP-REQ message and a BS responds with an MOB_SLP-RSP message. While these request and response messages are exchanged, the sleep mode parameters, such as initial-sleep window (T_{min}), final-sleep window (T_{max}), listening window, and start frame number, are negotiated. These parameters are used to decide the sleep interval and listening interval in each sleep cycle during PSC I operation.

Both MS and BS initiate the sleep mode operation at the promised start frame. The size of the first sleep window is set to the initial-sleep window T_{min}.

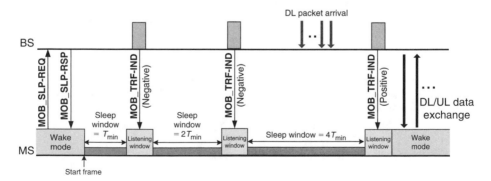

FIGURE 6.2
Operation of PSC I.

Then, the size of each sleep window increases binary exponentially every sleep but does not exceed the final-sleep window T_{max}. If the MS has reached T_{max}, it maintains the sleep window at T_{max}. That is, the size of sleep window in ith cycle is controlled by

$$T_i = \begin{cases} 2^{i-1}T_{min}, & \text{if } 2^{i-1}T_{min} < T_{max} \\ T_{max}, & \text{otherwise} \end{cases} \tag{6.1}$$

After each sleep interval the MS wakes up for a fixed-size listening interval, which generally has a short length. During each listening interval, the MS listens to the MOB_TRF-IND message that is broadcasted from the BS, which indicates whether any packets have arrived for the MS during the sleep interval. If this message delivers a positive indication, the MS exits sleep mode and enters wake mode to receive all the buffered packets from the BS. In addition, PSC I operation is finished when a BS transmits MAC data during any listening window or when the MS transmits a bandwidth request with respect to the connection belonging to the current PSC.

6.2.2 Power-Saving Class of Type II

PSC II is recommended for UGS and real-time variable rate (RT-VR) connections, such as VoIP and video-streaming. Figure 6.3 shows the basic sleep mode operation of PSC II. Similar to the case of PSC I, PSC II is activated by the exchange of MOB_SLP-REQ and MOB_SLP-RSP messages between an MS and a BS. For PSC II to work, it is necessary to set three parameters: initial-sleep window (T_{min}), listening window, and start frame number. Since real-time traffic is generated periodically, the sleep and listening windows in each sleep cycle have constant size. Therefore, the size of the sleep window is set to the initial-sleep window T_{min} at all times.

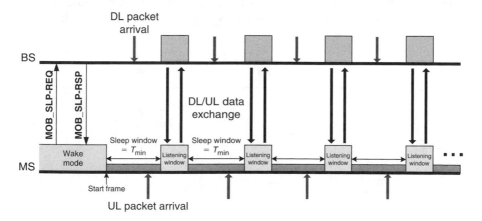

FIGURE 6.3
Operation of PSC II.

PSC II differs from PSC I in the following respects. During the listening interval, the BS does not transmit the traffic indication message (MOB_TRF-IND). Instead, the MS and BS exchange their real-time packets with each other directly. Hence, sleep mode is maintained uninterrupted, which is more efficient for real-time traffic because the signaling overhead required to restart sleep mode can be eliminated. In PSC II, sleep mode is terminated by the specified management message (i.e., MOB_SLP-REQ or MOB_SLP-RSP) issued by either the MS or BS.

6.2.3 Power-Saving Class of Type III

PSC III is recommended for multicast connections, as well as for management operations, such as periodic ranging, dynamic service operations, and advertisement message broadcasting. Two parameters, final-sleep window and start frame number, are required for PSC III. An MS using PSC III initiates sleep mode operation at the start frame number, and powers off during a sleep interval specified as the size of the final-sleep window. After the expiration of one sleep interval, the MS powers on and PSC III operation finishes automatically. PSC III allows just one sleep cycle at a time and terminates automatically unless another sleep request is made at the time that the final-sleep window is finished.

Figure 6.4 shows a basic PSC III operation used for the periodic ranging. PSC III can be activated efficiently by a next periodic ranging type/length/value (TLV) encoding included in an RNG-RSP message. If a next periodic ranging TLV encoding in a certain RNG-RSP message is set to a positive value during the periodic ranging process, it activates a special PSC III associated with the ranging process. If the MS confirms that the RNG-RSP message contains TLV encoding, it starts PSC III operation at the next frame immediately and continues to sleep during the frames that the next periodic ranging TLV indicates. When the MS's sleep period ends, the MS and BS perform a periodic ranging each other. After completing periodic ranging successfully, the BS instructs

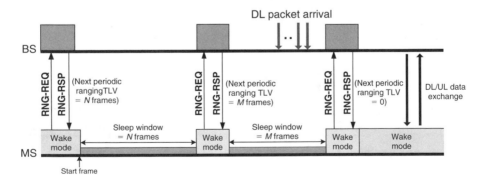

FIGURE 6.4
Operation of PSC III.

the MS to enter sleep mode again for the duration of the time indicated by the next periodic ranging TLV encoding in the RNG-RSP message. If downlink traffic addressed to the MS arrives during PSC III operation, the BS informs the MS of this fact by sending an RNG-RSP message that includes the next periodic ranging TLV with a value set to zero. If an MS receives the RNG-RSP message with this indication, it immediately deactivates PSC III and resumes normal operation with the BS to receive pending data.

6.3 Handover

The IEEE 802.16e system provides an HO function to support the mobility of MS. When the signal quality of the current BS worsens due to fading or interference due to mobility, the MS hands over to another BS that provides better signal quality and quality of service (QoS). The HO procedure in the IEEE 802.16e is mainly divided into two processes: network topology acquisition and HO operation. In addition, macro diversity HO (MDHO) and fast BS switching (FBSS) techniques are proposed as optional modes to support more seamless and faster HO.

6.3.1 Network Topology Acquisition

The object of the network topology acquisition is to collect information about a channel's description and its physical quality from an MS's neighboring BSs before an actual handover occurs. Information about the network topology is acquired by performing a network topology advertisement process and a scanning process. In addition, an MS can execute an association process during the scanning process, which is an optional initial ranging procedure formed between the MS and a target BS to which the MS wants to connect.

6.3.1.1 Network Topology Advertisement

A BS advertises information about the network topology by an MOB_NBR-ADV message, which is broadcasted periodically by the BS. It provides the number of neighboring BSs and channel information for each neighboring BS. It contains physical frequency, downlink channel descriptor (DCD), and uplink channel descriptor (UCD) messages according to each neighboring BS's identity (BSID).

According to the IEEE 802.16e standard, the BS should transmit one MOB_NBR-ADV message at least every 30 s. To make the MOB_NBR-ADV message, a serving BS gathers channel information about each neighboring BS over the backbone. If an MS receives this message, it knows how many BSs there are nearby and their channel information (i.e., DCD and UCD contents). This network topology information is used for the MS's scanning process and facilitates MS synchronization with neighboring BSs, because the MS does not have to monitor their DCD/UCD broadcasts.

6.3.1.2   Scanning of Neighbor BSs

Once an MS is made aware of the existence of neighboring BSs by reception of the MOB_NBR-ADV message, it monitors their suitability to find a target BS for HO; that is, the MS scans neighboring BSs. For this scanning process, the following messages are defined:

*   *MOB_SCN-REQ* is issued by the MS to request scanning and negotiate a number of scanning parameters, such as scan duration, interleaving interval, and the number of scan iteration.

*   *MOB_SCN-RSP* is sent by the BS as a response to the MOB_SCN-REQ message, to inform the MS whether it approves or rejects the scanning request. It contains the final scanning allocation parameters and the start frame number for initializing scanning.

*   *MOB_SCN-REP* is transmitted by the MS to report the scanning results, which can be carrier to interference noise ratio (CINR), received signal strength indication (RSSI), relative delay, or round trip delay (RTD). The MS can transmit this message to its serving BS at anytime or at the time indicated in the MOB_SCN-RSP message after each scanning period.

Figure 6.5 shows the operation of the network topology advertisement and scanning. First, an MS receives an MOB_NBR-ADV message and is informed of the existence of two neighboring BSs. If the trigger condition specified in the DCD information is satisfied, the MS sends its serving BS an MOB_SCN-REQ message to activate a scanning process. This request message contains the following scanning allocation parameters: the size of the scanning interval, the size of the interleaving interval, and the number of scan iterations. The scanning interval expresses a period during which the MS can scan for available BSs. The interleaving interval indicates a period during which the MS can operate normally and can receive/send data from/to its serving BS. The number of scan iterations determines how many times the scanning and interleaving intervals are repeated during the total scanning period. If the BS receives the MOB_SCN-REQ message, it responds with an MOB_SCN-RSP message. The MOB_SCN-RSP message can either grant a scanning interval that is at least as large as that which the MS requests or reject the scanning request by setting the value of scan duration to zero.

After receiving the MOB_SCN-RSP message that approves the scanning request, the MS starts to scan for neighboring BSs at the start frame and continues throughout the scanning interval specified in the response message. When a neighboring BS is identified by scanning, the MS attempts to synchronize with its downlink transmissions and estimates the quality of its physical channel. After the end of each scanning interval, the MS may issue the MOB_SCN-REP message to report the scanning results. The serving BS should buffer incoming data addressed to the MS during the scanning interval and forward those data after the scanning interval during any interleaving interval

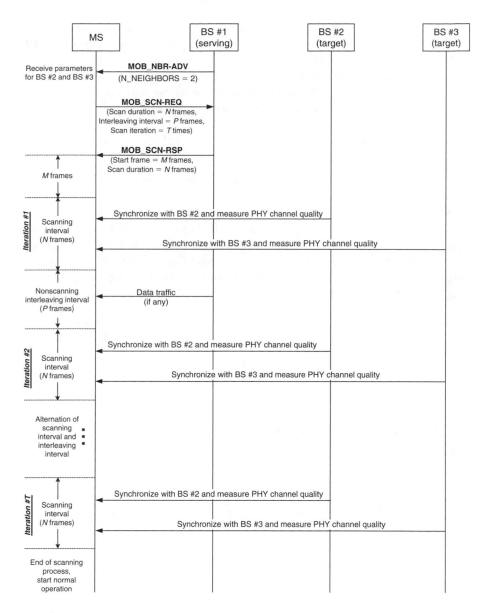

FIGURE 6.5
Operation of network topology advertisement and scanning.

or after completing the scanning operation. These scanning and interleaving intervals are repeated alternately for the number of scan iterations specified.

6.3.1.3 Association

Association is an optional initial ranging procedure performed with one of the neighboring BSs during the scanning interval. Association enables the

MS to acquire and record information about ranging parameters and service availability, for the purpose of properly selecting the HO target and expediting a potential future HO to a target BS. The recorded ranging parameters of an associated BS can be further used to set the initial ranging values in future ranging events during an actual HO. According to the use of BS coordination and network assistance, there are three levels of association:

- Association level 0: Scan/association without coordination
- Association level 1: Association with coordination
- Association level 2: Network-assisted association reporting

At association level 0, the MS performs the basic initial ranging process with each target BS during the scanning interval. Figure 6.6 shows the scanning operation at association level 0. The exchange of the MOB_SCN-REQ and MOB_SCN-RSP messages requires the process of association level 0 together with the scanning process. During the scanning interval, the MS not only scans but also performs the initial ranging process with the target BSs referred in

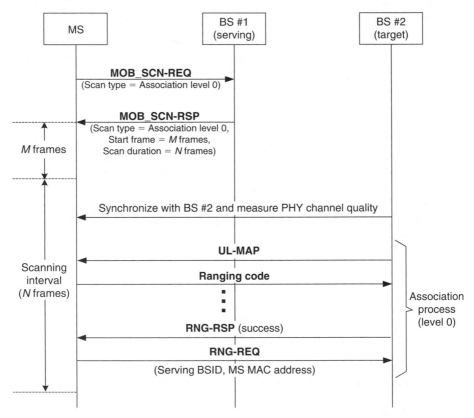

FIGURE 6.6
Operation of scanning with association of level 0.

the MOB_NBR-ADV message. Since the target BS has no knowledge of the MS and provides only contention-based ranging allocations, the MS chooses randomly a ranging code from the initial ranging domain of the target BS and transmits it in the contention-based ranging region. After the target BS has received the ranging code and sends an RNG-RSP message with the ranging status "success," it will provide an uplink allocation of adequate size for the MS to transmit an RNG-REQ message. Then, the MS transmits the RNG-REQ message with the serving BSID and its MAC address related to the association ranging. Association level 0 uses only a basic initial ranging process for the association with the target BS and does not require any coordination of its serving BS. However, this simplicity may cause collisions of ranging codes during the association process; hence, time required to complete the association process is increased.

At association level 1, the serving BS provides the MS with the association parameters and coordinates the association between the MS and neighboring BSs, to reduce the time required for association. Figure 6.7 shows the scanning operation at association level 1. At association level 1, each neighboring

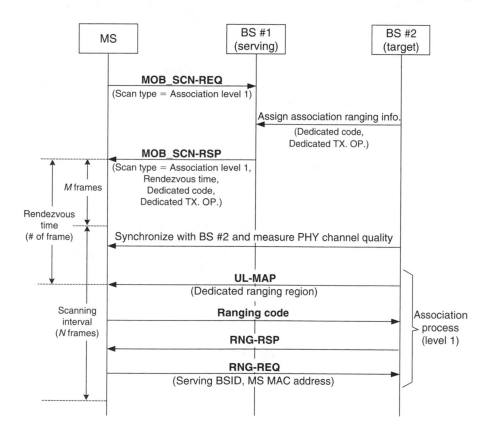

FIGURE 6.7
Operation of scanning with association of level 1.

BS provides the association parameters (ranging region, unique code number, and dedicated transmission opportunity) at a predefined rendezvous time. The serving BS informs the MS of these parameters by sending an MOB_SCN-RSP message. The rendezvous time specifies the frame in which the neighboring BS will transmit a UL-MAP containing the definition of the dedicated ranging region where the MS uses the assigned ranging code. The rendezvous time is defined as units of frames, which begins at the frame where the MOB_SCN-RSP message is transmitted. In the scanning interval, the MS synchronizes with the neighboring BS first, reads the UL-MAP transmitted at the rendezvous time, and extracts the description of the dedicated ranging region from this UL-MAP. Then, the MS determines the specific region where it should transmit the dedicated ranging code at the dedicated transmission opportunity. Neighboring BSs will assign a different code or a different transmission opportunity for the association, so there is no potential for transmissions from different MSs to collide. Hence, association will be fast.

Association level 2 is similar to association level 1. The difference is that an MS does not have to wait to receive the RNG-RSP from a neighboring BS after it transmits the ranging code to it. Instead, the RNG-RSP information is sent from each neighboring BS to the serving BS over the backbone. The serving BS aggregates all ranging information into a single MOB_ASC-REP message and transmits it to the MS. When receiving this report message, the MS updates its association database (physical offsets, time offsets, and channel identities (CIDs)) for each associated BS. Association level 2 supports fast association without access collision and the efficient reception of aggregating association information, but it requires more signaling overhead between the serving BS and target BSs.

6.3.2   Basic Handover Operation

HO is essential for supporting MS mobility in mobile cellular environments, and it enables an MS to change its air interface from one BS to another. Figure 6.8 illustrates a basic HO procedure in the IEEE 802.16e system. An HO observes the following procedures: (1) cell reselection, (2) HO decision and initiation, (3) HO cancellation, (4) synchronization to target BS downlink, (5) use of scanning and association results, (6) ranging, (7) termination with the serving BS, (8) drops during HO, and (9) network entry/reentry. The messages related with the HO process are as follows:

- *MOB_MSHO-REQ* is issued by an MS to initiate an HO. It contains the information about the recommended neighboring BSs.
- *MOB_BSHO-RSP* is sent by a BS in response to reception of the MOB_MSHO-REQ message. It delivers the information about the recommended neighboring BSs for HO.
- *MOB_BSHO-REQ* is issued by a BS that wants to initiate an HO. The MS receiving this message scans the recommended neighboring BSs specified in this message.

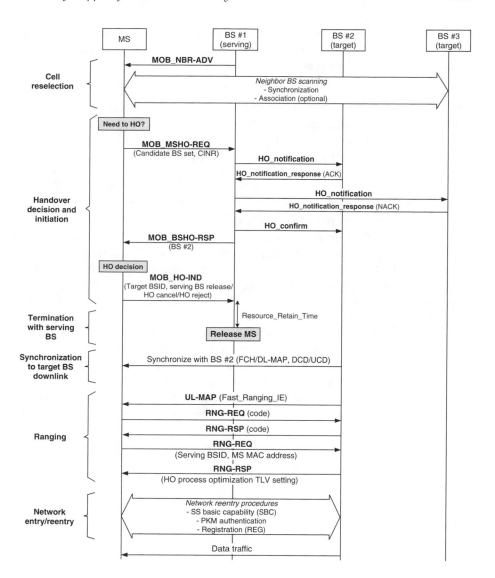

FIGURE 6.8
Handover operation.

- *MOB_HO-IND* is transmitted from the MS to its serving BS to inform of the final HO indication, which may result in serving BS release, HO cancellation, or HO rejection.

6.3.2.1 Cell Reselection

Cell reselection refers to the process of an MS scanning or association with one or more BSs to determine their availability and suitability as an HO target. To perform cell reselection, the MS uses the information acquired from an

MOB_NBR-ADV message and the serving BS's scheduled scanning intervals. Therefore, cell reselection process does not involve terminating an existing connection with a serving BS.

6.3.2.2   Handover Decision and Initiation

An HO is initiated by a decision to handover from a serving BS to a target BS. The decision originates at either an MS or a serving BS. The MS can initiate HO by transmitting an MOB_MSHO-REQ message. To acknowledge the MOB_MSHO-REQ, the BS responds with an MOB_BSHO-RSP message. The BS can initiate HO by sending an MOB_BSHO-REQ message in unsolicited manner. If the serving BS receives the MOB_MSHO-REQ message or judges that the MS needs to perform a HO, it sends an HO notification message containing the MS information to one or more potential target BSs over the backbone network, to notify that the MS intends to HO. If the serving BS receives an HO notification response from the target BSs, it selects a target BS suitable for the MS's HO according to the status of the response message (accept or reject), and then sends an HO confirm message to the selected target BS. Thereafter, the serving BS informs the MS of the selected target BS by sending the MS the MOB_BSHO-RSP message (in the case of MS-initiated HO) or the MOB_BSHO-REQ message (in the case of BS-initiated HO).

If the MS receives the MOB_BSHO-RSP or MOB_BSHO-REQ message, it makes a final HO decision and sends an MOB_HO-IND message. The MOB_HO-IND message notifies the serving BS of the final decision, which can be a serving BS release, HO cancellation, or HO rejection. If the BS receives the MOB_HO-IND with an option of serving BS release, it sets a resource retain timer. When the resource retain timer expires, the MS is disconnected from its serving BS and can no longer monitor downlink traffic from its serving BS.

6.3.2.3   Handover Cancellation

The MS can cancel the current HO at any time, regardless of whether it was the MS or BS that initiated the HO. This cancellation is made by transmitting the MOB_HO-IND with the HO cancel option. When the serving BS receives the MOB_HO-IND with the HO cancel option before the resource retain timer expires, the MS and serving BS resume normal communication. If an MS wants to attempt to handover to a different BS, whether or not that BS was included in MOB_BSHO-RSP or MOB_BSHO-REQ, it requests the serving BS to reject its current HO instruction by sending an MOB_HO-IND with the HO reject option. If the BS confirms this request, it reconfigures a list of neighboring BSs and retransmits the MOB_BSHO-RSP message, which will include a new list of neighboring BSs.

6.3.2.4   Synchronization to Target BS Downlink

To connect with the target BS, the MS synchronizes with the downlink transmissions of the target BS and obtains downlink (DL) and uplink (UL) transmission parameters. If the MS had previously received an MOB_NBR-ADV message including target BSID, physical frequency, DCD, and UCD, this synchronization process can be shortened. Otherwise, the MS synchronizes with the target BS by scanning the possible channels of DL frequency band until it finds a valid DL signal.

6.3.2.5 Use of Scanning and Association Results

An MS scans the target neighboring BSs and has the option to try association. If the target BS has previously received HO notification over the backbone from the serving BS, the target BS can place a fast ranging information element (Fast_Ranging_IE) in the UL-MAP to allocate a noncontention-based initial ranging opportunity. Therefore, the MS can use the noncontention-based initial ranging opportunity by scanning the UL-MAP of the target BS for fast HO ranging process.

6.3.2.6 Ranging

An MS and a target BS conduct an initial ranging or HO ranging after the synchronization with the target BS downlink. An MOB_BSHO-REQ or MOB_BSHO-RSP message informs the MS of the common time interval at which the dedicated initial ranging transmission opportunity for the MS will be provided by the target BS. Therefore, the MS can receive the Fast_Ranging_IE in the UL-MAP of its target BS, which includes a noncontention-based initial ranging opportunity. If the MS confirms that initial ranging opportunity, it can transmit an RNG-REQ code to the target BS without access collision. This operation enables fast ranging because the target BS provides a dedicated UL resource for the ranging request.

6.3.2.7 Termination with the Serving BS

If the MS decides to carry out an HO after receiving an MOB_BSHO-RSP or MOB_BSHO-REQ message, the MS terminates service with the serving BS. This operation is accomplished by sending an MOB_HO-IND message with the option of serving BS release. If the BS confirms the release of its service, it starts the resource retain timer. Until the resource retain timer expires, the serving BS retains the MS connections, MAC state machine, and packet data associated with the MS for service continuation. When the resource retain timer expires, the serving BS releases all information about the MS and the MS is disconnected from its serving BS. However, regardless of resource retain timer, the serving BS can remove the MAC context and MAC data associated with the MS if it receives a backbone message from the target BS that indicates that the MS is attached to the target BS over the network.

6.3.2.8 Drops during Handover

A drop occurs when an MS has ceased to communicate with its serving BS before the normal HO procedure has been completed. An MS can detect a drop by its failure to demodulate the DL, or by the failure of the periodic ranging mechanism. When the MS has detected a drop during network reentry with a target BS, it attempts network reentry with its preferred target BS by the cell reselection procedure. At this time, the MS can try to resume communication with the serving BS by sending an MOB_HO-IND message with the HO cancel option. If the MS fails to establish network reentry with its preferred target BS, the MS performs the initial entry procedure.

6.3.2.9   *Network Entry/Reentry*

An MS starts to perform network entry procedures with a new BS after a successful ranging process. If the MS has sent an RNG-REQ that includes a serving BSID during the ranging process, a target BS may request information about the MS from the serving BS over a backbone network, and the serving BS may respond with the requested information, to expedite the network entry process. Therefore, the process of network entry with the target BS can be shortened if the target BS obtains information about the MS from the original serving BS. Depending on the amount of that information, the target BS can decide to skip one or more steps among the following network entry procedures: negotiate basic capabilities, privacy key management (PKM) authentication phase, traffic encryption key (TEK) establishment phase, and registration. This HO optimization mechanism is an effective technique for reducing the time required for network entry in the IEEE 802.16e system.

6.3.3   Macro Diversity Handover and Fast BS Switching

In addition to the HO operation discussed above, there are two optional HO techniques: MDHO and FBSS. The purpose of both HO schemes is to provide a diversity gain that increases cell coverage and QoS at a cell boundary, as well as a fast HO.

MDHO performs the diversity combining both DL and UL, since two or more BSs transmit the same DL data to the MS and receive the same UL data from the MS in the same time interval. FBSS HO utilizes selection diversity and a fast switching mechanism to improve link quality. In FBSS, the MS only transmits/receives data to/from its serving BS (called the anchor BS when this technique is in operation) at any given frame. The anchor BS can change, frame by frame, according to a scheme for selecting BSs.

There are several requirements that enable MDHO and FBSS to occur between the MS and a group of BSs, as follows:

- The BSs involved in MDHO/FBSS are synchronized based on a common time source.
- The frames sent by the BSs involved in MDHO/FBSS at a given frame time arrive at the MS within a predetermined interval.
- The BSs involved in MDHO/FBSS have a synchronized frame structure.

- The BSs involved in MDHO/FBSS have the same frequency assignment.
- The BSs involved in MDHO/FBSS are required to share or transfer a MAC context.

In the case of MDHO, the following two conditions are required additionally:

- The BSs involved in MDHO use the same set of CIDs for the connections that are established with the MS.
- The same data are sent to the MS by all BSs involved in MDHO.

6.3.3.1 Macro Diversity Handover

In MDHO, an MS and a BS manage a diversity set, which is a list of BSs that are involved in MDHO with the MS. Among the BSs in the diversity set, an anchor BS is defined. When operating in MDHO, the MS can communicate with all BSs in the diversity set for UL and DL traffic. For DL MDHO, two or more BSs provide synchronized transmission of MS DL data such that the diversity combining can be achieved by the MS. For UL MDHO, the transmission from an MS is received by multiple BSs such that the selection diversity can be achieved by multiple BSs.

Figure 6.9 shows a procedure of MDHO. A BS that supports MDHO or FBSS broadcasts the DCD message that includes the H_Add and H_Delete thresholds. These thresholds are used by an MS with FBSS/MDHO capability to determine when the MOB_MSHO-REQ should be sent. When the long-term CINR of a neighboring BS is higher than the H_Add threshold, the MS sends the MOB_MSHO-REQ to require that this neighboring BS be added to a diversity set. When the long-term CINR of a serving BS is less than the H_Delete threshold, the MS sends the MOB_MSHO-REQ to require that this serving BS be removed from the diversity set.

In Figure 6.9, an MS that communicates with its serving BS (BS1) transmits an MOB_MSHO-REQ message if the CINR of a neighboring BS is higher than the H_Add threshold. The MOB_MSHO-REQ message contains not only a possible list of BSs to be included in the MS's diversity set, but also their channel quality evaluated using previous channel measurements. When sending an MOB_BSHO-RSP, the BS provides a list of BSs recommended for the MS's diversity set. In Figure 6.9, a BS2 is added into the diversity set. Moreover, the BSs can provide a recommended list of BSs by sending an MOB_BSHO-REQ in an unsolicited manner. If the MS receives the MOB_BSHO-RSP message, it chooses the actual update by considering the received diversity set and sends an MOB_HO-IND message that contains the type field of confirm diversity set update. Finally, the MS can receive DL-MAP/UL-MAP from BS2 as well as from BS1 (which is the anchor BS in the diversity set), so it can communicate with BS1 and BS2 simultaneously.

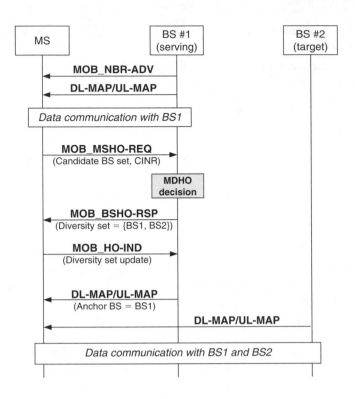

FIGURE 6.9
MDHO procedure.

The MS can reject the diversity set recommended by the anchor BS by setting the type field in MOB_HO-IND to diversity set update reject. In this case, the BS reconfigures the diversity set list and retransmits the MOB_BSHO-RSP message to the MS. In addition, the MS can cancel a diversity set update at any time during a diversity set update process. The cancellation is made by transmitting an MOB_HO-IND with the type field set to diversity set update cancel.

6.3.3.2 Fast BS Switching

The MS and the BS involved in FBSS manage a diversity set by using the same threshold mechanism and an anchor BS is defined in the diversity set. When operating in FBSS, the MS only communicates with the anchor BS for UL and DL data. The transition from one anchor BS to another BS is performed without invoking the normal HO procedure. The FBSS procedure is shown in Figure 6.10. Anchor BS updating begins when the MS sends an MOB_MSHO-REQ or the anchor BS sends an MOB_BSHO-REQ. The preferred anchor BS is a member of the MS's current diversity set. The MS selects the preferred anchor BS through a prior measurement of signal strength and reports it to the serving BS by using the MOB_MSHO-REQ message. A BS decides the target anchor BS

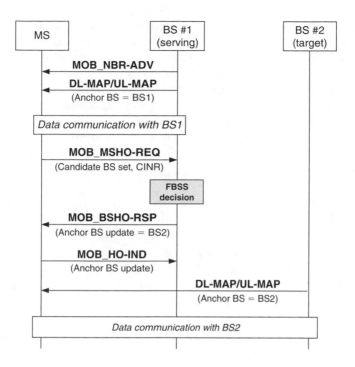

FIGURE 6.10
FBSS procedure.

on the basis of the MS report and then informs the MS of the anchor BS update by sending an MOB_BSHO-RSP containing the estimated switching time. Figure 6.10 illustrates a case in which the anchor BS is updated from BS1 to BS2. The MS updates its anchor BS on the basis of the information received in the MOB_BSHO-RSP message. The MS indicates its acceptance of the new anchor BS by sending an MOB_HO-IND message with the type field set to confirm anchor BS update. At this time, the MS can receive data from a new anchor BS (BS2). The MS can also reject or cancel the anchor BS update instruction by sending an MOB_HO-IND message with the type field set to reject or cancel.

6.4 Paging and Location Update

The IEEE 802.16e system defines an MS idle mode to provide paging and location update mechanisms. The MS can be in idle mode when there is no traffic to/from the MS for a given period. Idle mode allows an MS to become periodically available for DL broadcast traffic messaging without registering with a specific BS while it traverses an air link environment consisting of multiple BSs. An MS in idle mode does not have to perform HO and can suspend all normal operation requirements. Hence, it can conserve power and operational resources, and the network can eliminate air interface and HO traffic. However, for MSs in idle mode, a network and BS broadcasts or multicasts a paging message periodically in the paging area to inform the MS of its pending DL traffic, and the MS should scan for the paging message in every discrete interval and inform the network of its current location.

Several BSs compose a logical group called a paging group, the purpose of which is to offer a contiguous coverage region in which the MS checks only the DL paging message to see whether there is traffic targeted to it. The paging groups are defined and managed by the management system (e.g., paging controller) in the network. For idle mode, the following messages are defined:

- *DREG-REQ* is sent by the MS to request deregistration from its serving BS or initiation of idle mode.
- *DREG-CMD* is transmitted by the BS to force the MS to change its state. The BS can transmit the DREG-CMD in an unsolicited manner or as a response to the DREG-REQ message. Upon receiving a DREG-CMD, the MS performs the action indicated by this command message.
- *MOB_PAG-ADV* is broadcasted or multicasted by the BS during the paging interval. This message requests the MS to update its location or reenter the network.

6.4.1   Basic Paging Operation

Paging begins after the MS deregisters. Figure 6.11 illustrates the basic paging operation. First, an MS in active mode sends a DREG-REQ to request deregistration and enters idle mode. If the BS receives the DREG-REQ, it sends a DREG-CMD message to the MS. A serving BS may also induce an MS to enter idle mode by sending an unsolicited DREG-CMD message. Upon receipt of the unsolicited DREG-CMD message from the serving BS, the MS sends a DREG-REQ message and then enters idle mode.

In idle mode, the MS and BS release all connections, all air resources, and IP address, but the serving BS or the paging controller that administers idle mode activity for the MS can retain certain MS services and operational information, which it can use to expedite a future network reentry from idle mode on the part of the MS. For idle mode operation, the MS maintains an idle mode timer and the paging controller maintains an idle mode system timer. These two timers are set to the same value and start when the serving BS transmits the DREG-CMD message that directs the MS to enter idle mode, and recycle whenever the MS updates its location successfully while in idle mode. These two timers provide a time interval during which the MS should update its location so that it can be found in the network managed by the current paging controller. If the idle mode system timer has expired or if the MS enters/reenters the network and resumes normal operation, the paging controller discards all MS services and operational information retained for

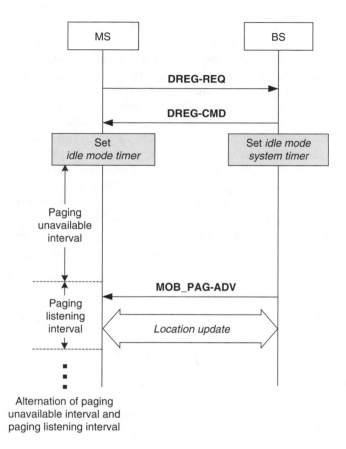

FIGURE 6.11
Paging operation.

idle mode management purposes. If the idle mode timer has expired, the MS should reenter the network, because the paging controller has discarded all MS information.

When the MS initiates idle mode, it selects a preferred BS, which can be a current serving BS or the neighboring BS that has the best air interface DL properties. The MS synchronizes and decodes the DCD and DL-MAP from the preferred BS to extract the frame size and current frame number. The MS uses these to determine the time interval between the present and the next regular paging time from the preferred BS. This calculated time interval becomes an MS paging unavailable interval. During this interval, the MS can power down, scan neighboring BSs, reselect a preferred BS, conduct the ranging, or perform other activities for which the MS will not guarantee availability to any BS for DL traffic. At the end of the MS paging unavailable interval, an MS paging listening interval starts. During this interval, the MS receives an MOB_PAG-ADV message broadcasted by the BS. The MOB_PAG-ADV is a notification message for MSs in idle mode, which indicates the presence of DL traffic pending or requests a location update. The paging listening interval has a frame unit of constant size and is repeated every paging cycle. After a paging listening interval, another paging unavailable interval begins. That is, the paging unavailable interval and paging listening interval are repeated alternately when the MS is in idle mode.

Idle mode is terminated when an MS reenters the network; when the paging controller fails to receive a response to paging messages and so realizes that the MS is unavailable, or when the idle mode system timer has expired. An MS terminates idle mode and reenters the network if it decodes an MOB_PAG-ADV message that contains an action code of enter network. In the event that an MOB_PAG-ADV message contains an action code of perform ranging, the MS updates its location to the network. In both cases, ranging code and transmission opportunities are assigned to the MS in the MOB_PAG-ADV message, so the MS can reenter the network or update its location by using the dedicated code and transmission opportunity without access collision.

6.4.2   Location Update

An MS in idle mode updates its location in the following circumstances:

- *Paging group update*: The MS updates its location when it detects a change in paging group. If the paging group identifier contained in an MOB_PAG-ADV broadcast message during the MS paging listening interval does not match the paging group to which the MS belongs, the MS determines that the paging group has changed.

- *Timer update*: The MS periodically updates its location prior to the expiration of an idle mode timer.

- *Power down update*: The MS attempts to update its location once as a part of its orderly power-down procedure. This mechanism enables a paging controller to update the MS's exact status and to delete all information about the MS and discontinue idle mode paging control for the MS at the time of power down.

- *MAC hash skip threshold update*: The MS updates its location when the MS MAC hash skip counter exceeds the MAC hash skip threshold successively. After successful location update, the BS and MS reinitialize their respective MAC hash skip counters.

Figure 6.12 illustrates location update. If an MS in idle mode decides to update its location, it attempts to update with a target BS. Location is updated by the exchange of RNG-REQ and REG-RSP messages. The MS sends an RNG-REQ message, which includes the ranging purpose indication of location update request and the paging controller ID. The target BS replies with an RNG-RSP message, which includes the location update response and paging group ID. If the location update is successful, the target BS notifies the paging controller of the location of the MS via the backbone, and the MS records the

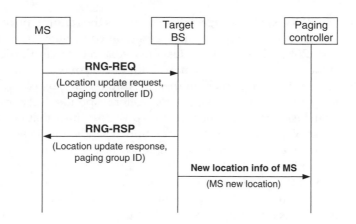

FIGURE 6.12
Location update procedure.

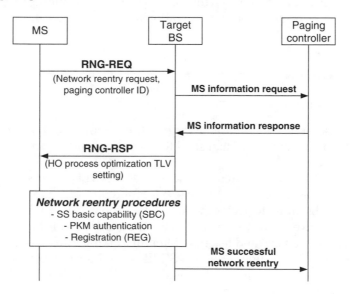

FIGURE 6.13
Procedure of network reentry from idle mode.

paging group ID of the target BS. In addition, the paging controller can send a backbone message to the BS at which the MS entered idle mode, to give notice that the MS has transferred to a different paging group.

6.4.3 Network Reentry from Idle Mode

An MS in idle mode reenters the network when it wants to connect to its network to receive/transmit data. Figure 6.13 shows the procedure of network reentry from idle mode. The MS initiates network reentry with the target BS by sending an RNG-REQ message, which includes the ranging purpose indication of network reentry request and the paging controller ID. If the target BS receives an RNG-REQ that includes a network reentry indication and it had not previously received information about the MS over the backbone, the target BS requests information about the MS from the paging controller over the backbone network, and the paging controller responds. Network reentry procedures can be shortened if the target BS possesses information about the MS. If the target BS possesses such information, it sends the MS an RNG-RSP with an HO process optimization TLV that indicates which reentry management messages can be omitted. Then, the MS and BS communicate with each other to perform the network reentry procedure with respect to such matters as the negotiation of SS basic capability, authentication, and registration. After the network reentry process is completed, normal operation is resumed. The target BS notifies the paging controller of the successful network reentry of the MS via the backbone, and the paging controller can also send a backbone message to the BS at which the MS entered into idle mode, to give notice that the MS has resumed normal operation at the new BS.

6.5 Summary

We have discussed the main mobility functions defined in the IEEE 802.16e standard: power-saving mechanism, HO operation, and paging and location update.

First, the IEEE 802.16e system provides three types of power-saving mechanisms. PSC I is used for nonreal-time services and provides a truncated binary exponential algorithm to decide the size of the sleep window, which is suitable for services with burst traffic attribute. PSC II is used for real-time services and provides periodic sleep and listening intervals, taking into account the traffic characteristics of real-time services. PSC III is used for multicast or management message transmission and provides an efficient sleep mechanism for aperiodic and continuous services.

Second, the IEEE 802.16e system provides a basic HO operation and enhanced mechanisms for fast and seamless HO. Network topology acquisition makes it possible for an MS to acquire information about the properties and quality of a channel from neighboring BSs before an actual handover. To obtain information about the network topology, the MS receives a network topology advertisement message from its serving BS and conducts scanning process and optional association with its neighboring BS. Basic HO operation is performed in the sequence cell reselection, HO decision and initiation, synchronization to target BS downlink, ranging, termination with a serving BS, and network entry. For a smooth HO, the sequential signaling procedure among the MS, serving BS, and target BSs is performed and an accurate decision algorithm is required. In addition, MDHO and FBSS support a fast and seamless HO, because they enable diversity combining and soft HO. However, this operation places many requirements on the BS and MS.

Third, the IEEE 802.16e system provides paging and location update operations. Paging mechanism allows an MS to operate in idle mode. The MS only updates its location and checks a paging message periodically when in idle mode. This mechanism offers advantages with respect to an MS's energy conservation and the reduction of used radio resources. Location update should be performed between an MS and a target BS to manage the location of the MS during idle mode operation. Network reentry is conducted when an MS wants to exit from idle mode. Network reentry follows a general network entry procedure, following which the MS can operate normally with a BS.

References

1. IEEE Std 802.16e-2005, *Part 16: Air Interface for Fixed and Mobile Broadband Wireless Access Systems*, Feb. 2006.
2. IEEE Std 802.16-2004, *Part 16: Air Interface for Fixed Broadband Wireless Access Systems*, Oct. 2004.
3. T. Kwon, H. Lee, S, Choi, J. Kim, D.-H. Cho, S. Yun, W.-H. Park, and K.-H. Kim, Design and implementation of a simulator based on a cross-layer protocol between MAC and PHY layers in a WiBro Compatible IEEE 802.16e OFDMA system, *IEEE Communications Magazine*, Vol. 43, Issue 12, pp. 136–146, Dec. 2005.
4. A.K. Salkintzis and C. Chamzas, Performance analysis of a downlink MAC protocol with power-saving support, *IEEE Transactions on Vehicular Technology*, Vol. 49, Issue 3, pp. 1029–1040, May 2000.
5. S. Choi, G.-H. Hwang, T. Kwon, A.-R. Lim, and D.-H. Cho, Fast handover scheme for real-time downlink services in IEEE 802.16e BWA system, *Vehicular Technology Conference 2005 Spring*, Vol. 3, pp. 2028–2032, May 2005.
6. X. Wu, B. Mukherjee, and B. Bhargava, A crossing-tier location update/paging scheme in hierarchical cellular networks, *IEEE Transactions on Wireless Communications*, Vol. 5, Issue 4, pp. 839–848, Apr. 2006.
7. Y. Zhang and M. Fujise, Energy management in the IEEE 802.16e MAC, *IEEE Communications Letters*, Vol. 10, Issue 4, pp. 311–313, Apr. 2006.

7

Measured Signal-Aware Mechanism for Fast Handover in WiMAX Networks

Jenhui Chen and Chih-Chieh Wang

CONTENTS

7.1 Introduction

Voice over Internet protocol (VoIP) has been established in the workplace as a transport mechanism for both fixed and wireless infrastructures. Switching voice paths within the existing packet-switched data networks as IP packets means that there is no need for separating voice and data infrastructures, and the traditional private branch exchange (PBX) can be replaced by a single server capable of supporting thousands of IP handsets. These devices look like regular phones but are handled more like personal computers (PCs), carrying their own unique identities with them wherever they connect to the network.

With the demand for wireless access and high bandwidth transmissions, fixed broadband wireless access (BWA) systems such as the local multipoint distribution service (LMDS) are proposed to provide multimedia services to a number of discrete subscriber sites with IP and offer numerous advantages over wired IP networks. This is accomplished by using base stations (BSs) to provide network access services to subscriber sites based on IEEE 802.16 WirelessMAN standard [11]. The progress of the standard has been fostered by the keen interest of the wireless broadband industry to capture the emerging worldwide interoperability for microwave access (WiMAX) market, the next-wave wireless market that aims to provide wireless broadband Internet services. The WiMAX Forum, formed in 2003, is promoting the commercialization of IEEE 802.16 and the European Telecommunications Standard Institute's (ETSI's) high-performance radio metropolitan area networks (MANs) (HyperMANs). It provides one of the potential solutions to beyond third generation/4th generation (B3G/4G) architecture [19,22].

IEEE 802.16e standard [16] provides a series of handover procedures for supporting mobility in BWA networks. Three different handover levels of association—Level 0 (L_0), Level 1 (L_1), and Level 2 (L_2)—are investigated for supporting roaming in the WiMAX network. The minimum required handover processing time (also known as service disruption time (DT)) of each levels are evaluated in Ref. 9 and are 280, 230 and 60 ms, respectively. Banerjee and his coauthors [3] analyzed and concluded that a DT of 50 ms is sufficient for media streams, while an interruption of 200 ms is generally acceptable. Meanwhile, it also showed that a DT of 500 ms will cause a perceptible interruption, which is unacceptable. Hence the present version of IEEE 802.16e is not sufficient for delay-sensitive applications, such as VoIP and video conference, since it will encounter a long handover processing delay due to its long ranging process, reassociation, reauthorization, and network transmission delay.

One feasible solution (to overcome this drawback) to conspicuously reduce the handover delay time is to proportionally reduce the number of forward-and-back turnaround times. Besides, many other methods were proposed to fulfill this goal in literature. Some of them focused on optimizing the cutoff parameters and appropriate queue sizes that minimize the overall blocking probability as handover occurs, such as the measurement-based priority scheme (MBPS) [24] and the signal prediction priority queueing (SPPQ) [5]. Also, some researches proposed using special or dedicated channels for handover calls, such as guard channel method (GCM) [15]. These methods will significantly reduce the handover failure probability and hence improve the handover performance. In addition, owing to the mobility and fading channel effect, the received signal strength (RSS) will vary with time and dynamically change following various environment conditions. Xhafa and Tonguz [25] demonstrated an analytical framework of handover to analyze the dynamic handover failure probability and estimated the order of handover calls to raise the successful probability of a handover.

Nevertheless, none of the above-mentioned schemes deal with the mechanism that preassigns a channel to a mobile subscriber station (MSS) for handover according to the movement of the MSS. Assume that a serving base station (SBS) knows the exact position of the MSS, the SBS could coordinate with the neighboring BSs (nBSs) around the MSS for handover preparation if the MSS appears in the boundary among the nBSs. The position information of the MSS and its corresponding movement intention could be estimated by observing the moving history of the MSS in recent records. There have been many measured mechanisms proposed for location management in general [1,6,8,20], which studied random mobility model for mobility estimation in wireless networks. Although the above-mentioned mechanism can enhance the successful probability of handover call, none of them aim at speeding up the handover processing time. Thus, in this chapter we will describe how to use measured signal-aware mechanism to aid speeding up the handover procedures. This mechanism can help the WiMAX system to support VoIP in high-speed mobility environment.

7.2 Legacy IEEE 802.16e Handover Procedures

To begin with the introduction of the proposed mechanism, we first review the architecture of the legacy IEEE 802.16e standard. The architecture of IEEE 802.16e is based on the Internet connecting several BSs through wired package-switched network as shown in Figure 7.1(1). An MSS communicates with a BS in an active set by using WiMAX technology through the air interface as shown in Figure 7.1(2).

Association of handover is an optional initial ranging procedure, which can be selected by the MSS. There are three handover levels of associations—L_0, L_1, and L_2—in IEEE 802.16e standard.

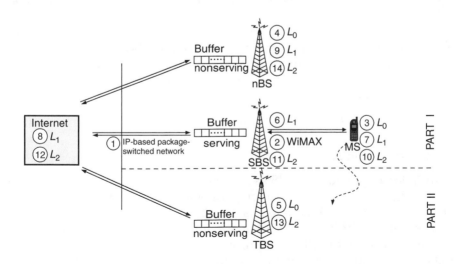

FIGURE 7.1
The architecture of IEEE 802.16e.

7.2.1 Association L_0

Association L_0 (scan and associate without coordination) is a contention-based ranging without coordination of handover. In general, the SBS allocates periodic intervals to MSS as shown in Figure 7.1(3). Therefore, the MSS may choose a ranging code arbitrarily to perform the initial ranging to all nBS as shown in Figure 7.1(4), which may include target base station (TBS) as shown in Figure 7.1(5). This ranging code is a contention-based resolution, which is based on the random backoff mechanism with an initial backoff countdown interval $CW = 2^{n+5}$, $n \in [0, 5]$ and the the maximum contention window size $CW_{max} = 1024$.

After the TBS successfully receives ranging code and sends ranging response (RNG-RSP) message with ranging status success, it will provide uplink allocation of adequate size for the MSS to transmit ranging request (RNG-REQ) message with type-length-value (TLV) parameters (SBS ID, MSS MAC address) related to the association ranging. The average time required for RNG-REQ message is denoted as T_{RNG}. In all cases, the MSS should synchronize with the new downlink first after the connection has been released by the SBS. The average time required to frame synchronization with the new downlink is denoted as T_{SYN}. The average time required during handover for reauthorization is denoted as T_{RA} (full authentication is assumed, where only 3-way handshake is performed instead of full authentication). The average time required for reregistration during handover is denoted as T_{RR}. The average time required for contention-based ranging (T_{CR}) process can be expressed as

$$T_{CR} = T_1 P_s + T_2 P_s(1 - P_s) + \cdots + T_n P_s(1 - P_s)^{n-1} \tag{7.1}$$

where T_n represents the mean contention window of the nth ranging and $T_n = CW_n/2, n \in [0, 5]$. P_s is the successful ranging probability, which can be calculated by $S = (CW - 1/CW)^N$, where N is the number of MSS. The service disruption time is defined as starting from the time the SBS or MSS sends a handover request to the time the MSS completes frame synchronization with the TBS. Therefore, we can get T_{L_0}, the service DT for L_0 scheme during the handover process by

$$T_{L_0} = T_{CR} + T_{RA} + T_{RNG} + T_{RR} + T_{SYN} \tag{7.2}$$

7.2.2 Association L_1

Association L_1 provides the MSS's association with coordination. In association L_1, the SBS provides association parameters to the MSS as shown in Figure 7.1(6 and 7)—Part I. $T_{SBS \leftrightarrows MSS}$ is the average transmission time required between SBS and MSS. The MSS may request to perform association with coordination by sending the scanning interval allocation request (MOB_SCN-REQ) message to the SBS with scanning type = 0b010. The SBS may also arrange for this type of association unilaterally by sending unsolicited the scanning interval allocation response (MOB_SCN-RSP) message. The SBS will then coordinate the association procedure with the requested neighboring BSs over the backbone as shown in Figure 7.1(6, 8, and 9). The average transmission time required between SBS and nBS is denoted as $T_{\text{SBS}\leftrightarrows\text{nBS}}$. T_{ID} is the average time required to Internet delay. Each neighboring BS may assign the same code or transmission opportunity to more than one MSS, but not both. Then, the MSS will synchronize to the neighbor BS as shown in Figure 7.1(7 and 9). The first frame immediately followed by the rendezvous time is denoted as T_R, including the uplink map (UL-MAP) transmitted time. The typical rendezvous time is between 0 and 500 ms [18]. Afterwards, the handover process will spend handover time T_{RA}, T_{RR}, and extract the description of the dedicated ranging region which will be set to 1. Therefore, we can get T_{L_1}, the service DT for L_1 scheme during the handover process by

$$T_{L_1} = T_{\text{MSS}\leftrightarrows\text{SBS}} + T_{\text{SBS}\leftrightarrows\text{nBS}} + T_R + T_{\text{RA}} + T_{\text{RR}} + T_{\text{ID}} + T_{\text{SYN}} \qquad (7.3)$$

7.2.3 Association L_2

Association L_2 (network-assisted association reporting) is indicated in Figure 7.1(10 and 11). The MSS may request to perform association with network-assisted association reporting by sending the MOB_SCN-REQ message, which includes the MSS-selected TBS, to the SBS with scanning type $= 0b011$. Then the SBS, as shown in Figure 7.1 (12 through 14), should request the TBS and the nBS with network-assisted association by sending the MOB_SCN-RSP message. The SBS will then coordinate the association procedure with the requested nBSs over the backbone as shown in Figure 7.1(11 through 14). The SBS may aggregate all ranging-related information into a single association result report (MOB_ASC-REP) message, which is called the RNG-RSP information. Afterward, the MSS is required to only transmit the code division multiple access (CDMA) ranging code at TBS as shown in Figure 7.1(10 and 13). When receiving this message, the MSS updates its association database (PHY offsets and CIDs) and timers for TBS. We can get T_{L_2}, the service DT for L_2 scheme during the handover process by

$$T_{L_2} = T_{\text{MSS}\leftrightarrows\text{SBS}} + T_{\text{SBS}\leftrightarrows\text{TBS}} + T_{\text{ID}} + T_{\text{SYN}} \qquad (7.4)$$

Table 7.1 is the comparison between the procedures of L_0, L_1, L_2 and the predicted handover scheme (PHS). From the table it is obvious how simple the PHS is.

TABLE 7.1

Comparison of L_0, L_1, L_2, and PHS

Scheme	Ranging	MSS⇆SBS	SBS⇆nBS	SBS⇆TBS
L_0	Contention-based	No	No	No
L_1	Limited ranging	⇆	⇆	No
L_2	Fast ranging	⇆	No	⇆
PHS	Fast ranging	←	No	No

7.3   Measured Signal-Aware Mechanism

The power received from a transmitter at separation distance d will directly impact the received signal-to-noise ratio (SNR). The desired signal level is represented as received power P_r in milliwatt and is given by

$$P_r \, [\text{mW}] = \frac{P_t G_t G_r}{\text{PL}(d)L} \quad [\text{valid if } d \gg 2D^2/\lambda] \tag{7.5}$$

where P_t is the transmitted power, G_t and G_r are the transmitter and receiver antenna gains, $\text{PL}(d)$ is the path loss (PL) with distance d, L the system loss factor ($L \geq 1$, transmission lines, etc., but not due to propagation), D the maximum dimension of transmitting antenna, and λ the corresponding wavelength of the propagating signal [23]. The antenna gain $G = 4\pi A_e/\lambda^2$; A_e is the effective aperture of the antenna. The length of λ can be obtained by $c/f = 3 \times 10^8/f$ in meters, where f is the frequency the signal carries. Besides, P_r can be represented in dBm units as

$$P_r \, [\text{dBm}] = 10 \log(P_r \, [\text{mW}])$$
$$= P_t + G_t + G_r - \text{PL}(d) - L \tag{7.6}$$

In the free space propagation model, the propagation condition is assumed idle and there is only one clear line-of-sight (LOS) path between the transmitter and receiver (T-R). On unobstructed LOS path between T-R, $\text{PL}(d)$ can be evaluated as $(4\pi)^2 d^2/\lambda^2$ or when powers are measured in dBm units as $92.4 + 20 \log(f) + 20 \log(d)$. We can get the desired T-R separation distance in meters

$$d = \frac{\lambda}{4\pi} \sqrt{\text{PL}(d)} = \frac{c}{4\pi f} \sqrt{\text{PL}(d)} \tag{7.7}$$

However, in street canyon scenario or urban environment, the PL model can be demonstrated through measurements using the parameter σ to denote the rule between distance and received power [2] and be expressed as

$$\text{PL}(d) = \text{PL}(d_0) + 10\rho \log\left(\frac{d}{d_0}\right) + X_\sigma + C_f + C_H \tag{7.8}$$

where the term $\text{PL}(d_0)$ is for the free space PL with a known selection in reference distance d_0, which is in the far field of the transmitting antenna (typically 1 km for large urban mobile systems, 100 m for microcell systems, and 1 m for indoor systems) and measured by $\text{PL}(d_0) = 20 \log(4\pi d_0/\lambda)$. The term X_σ denotes a zero-mean Gaussian distributed random variable (with units in dB) that reflects the variation in an average received power, which naturally occurs when PL model of this type is used [13]. The ρ is the path loss exponent, where $\rho = 2$ for free space and is generally higher for wireless channels.

It can be measured as $\rho = (a - bH_b + c/H_b)$, where a, b, and c are constants for each terrain category. The numerical values for these constants are studied in Ref. 12, where H_b is the height of the base station and $10\,\text{m} \leq Hb \leq 80\,\text{m}$. The term C_f, which is measured by $C_f = 6 \log (f/1900)$ [10], stands for the frequency correction factor; it accounts for a change in diffraction loss for different frequencies. Owing to the diffraction loss, a C_f is a simple frequency dependent factor. C_H is the receiver antenna height correction factor and H the receiver antenna height. $C_H = -10.7 \log(H/2)$ when $2\,\text{m} \leq H \leq 8\,\text{m}$. This correction factor closely matches the Hata–Okumura mobile antenna height correction factor for a large city [14].

We know that the audio or video quality of a receiver is directly related to the SNR. The limiting factor on a wireless link is the SNR required by the receiver for useful reception

$$\text{SNR [dB]} = P_r \, [\text{dBm}] - N_0 \, [\text{dBm}] \tag{7.9}$$

where N_0 is the noise power in dBm. Assuming the carrier bandwidth is B, the receiver noise figure F, the spectral efficiency r_b/B, and the coding gain G_c, the SNR for coded modulation with data rate r_b can be obtained by

$$\text{SNR [dB]} = 10 \log \left(\frac{P_r}{N_0} \frac{r_b}{B} \right) - G_c \tag{7.10}$$

where $N_0 \, [\text{dBm}] = -174 \, [\text{dBm}] + 10 \log B + F \, [\text{dB}]$. To obtain a criterion measurement of the received SNR, we force each MSS to use the lowest frequency to contend the channel with a predefined transmission power. The BS, after receiving a RNG-REQ message from the MSS, calculates the estimated distance between BS and MSS according to the received SNR. Assume that the BS needs a minimum receiving power or sensitivity $P_{r,\min}$, which corresponds to a minimum required SNR, denoted as $\text{SNR}_{\min}$, from each MSS to successfully receive the signal. According to Equations 7.6 and 7.10, we have

$$\begin{aligned} \text{SNR}_{\min} &= P_{r,\min} - N_0 \\ &= P_t + G_t + G_r - \text{PL}(d) - L - N_0 \end{aligned} \tag{7.11}$$

Substituting Equation 7.8 in Equation 7.11 leads to

$$\begin{aligned} \text{SNR}_{\min} = P_t + G_t + G_r - 20 \log \left(\frac{4\pi d_0 f}{c} \right) - 10\rho \log \left(\frac{d}{d_0} \right) \\ - X_\sigma - C_f - C_H - L - N_0 \end{aligned} \tag{7.12}$$

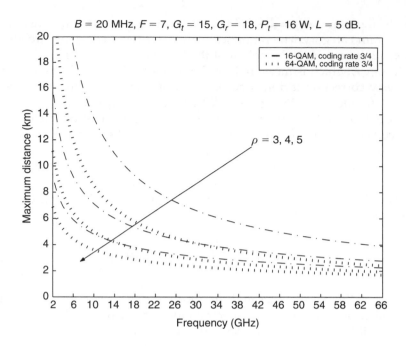

FIGURE 7.2
Maximum transmission distance versus frequency domains from 2 to 66 GHz in OFDM with different modulation schemes.

Solving Equation 7.12 for maximum transmission distance d denoted as d_{max}, we obtain

$$d_{max} = d_0 \times 10 \exp \left\{ \left[P_t + G_t + G_r - 20 \log \left(\frac{4\pi d_0 f}{c} \right) \right. \right.$$

$$\left. \left. - X_\sigma - C_f - C_H - L - \mathrm{SNR}_{min} - N_0 \right] \Big/ 10\rho \right\} \quad (7.13)$$

Figure 7.2, derived from Equation 7.13, shows the relation of the frequency and the distance between two isotropic antennas with different modulation schemes when the modulation is 16-QAM and 64-QAM and the required SNR_{min} is 18.2 dB and 22.4 dB, respectively.

7.4 Mobility Prediction

In this section, we will discuss the mobility of the MSS in detail. To prevent the out-of-service effect of MSSs due to mobility, we investigate a location prediction scheme to add to the PHS for channel migration. The IEEE 802.16e

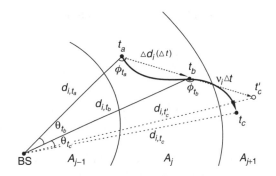

FIGURE 7.3
An illustration of mobility.

standard [16] recommends that the BS has to broadcast a REP-REQ message to all MSSs for channel measurements within 10 s to check whether the MSS is still in the service set. Therefore, the BS can get the SNR value by the replied REP-RSP message from each MSS to estimate the distance periodically.

Thus, as shown in Figure 7.3, the movement distance between time t_a and t_b of MSS$_i$ denoted as $\Delta d_i(\Delta t)$ can be calculated by using cosine theorem as

$$\Delta d_i(\Delta t) = \sqrt{d_{i,t_a}^2 + d_{i,t_b}^2 - 2 d_{i,t_a} d_{i,t_b} \cos \theta_{t_b}} \tag{7.14}$$

where θ_{t_b} can be estimated by using smart antenna systems [17,21] that employ antenna arrays coupled with adaptive signal-processing techniques at the BS. From Equation 7.14, the average velocity v_i of the MSS$_i$ is given by $v_i = \Delta d_i(\Delta t)/\Delta t = \Delta d_i(\Delta t)/(t_b - t_a)$.

To predict the maximum distance between the MSS$_i$ and the BS in time t_c denoted as t'_c, where $t'_c = t_b + \Delta t$, we have to obtain the ϕ_{t_a}. According to the cosine theorem, ϕ_{t_a} is obtained by

$$\phi_{t_a} = \cos^{-1} \left\{ \frac{d_{i,t_a}^2 + [\Delta d_i(\Delta t)]^2 - d_{i,t_b}^2}{2 d_{i,t_a} \Delta d_i(\Delta t)} \right\} \tag{7.15}$$

We simply suppose that each MSS moves forward directly. Then the moving distance can be estimated as $\Delta d'(t_c - t_b) = \Delta d(\Delta t) = v_i \Delta t$. Therefore, the estimated distance at time t'_3 will be

$$d_{i,t'_c} = \sqrt{d_{i,t_a}^2 + [\Delta d_i(\Delta t) + v_i \Delta t]^2}$$

$$- \sqrt{2 d_{i,t_a} [\Delta d_i(\Delta t) + v_i \Delta t] \cos \phi_{t_a}} \tag{7.16}$$

Substituting Equation 7.14 in Equation 7.16 we have

$$d_{i,t'_c} = \left[d_{i,t_b}^2 + 2v_i \Delta t \sqrt{d_{i,t_a}^2 + d_{i,t_b}^2 - 2d_{i,t_a}d_{i,t_b}\cos\theta_{t_b}} \right.$$
$$\left. - \frac{2v_i \Delta t d_{i,t_a}(d_{i,t_a} - d_{i,t_b}\cos\theta_{t_b})}{(d_{i,t_a}^2 + d_{i,t_b}^2 - 2d_{i,t_a}d_{i,t_b}\cos\theta_{t_b})^{1/2}} + (v_i \Delta t)^2 \right]^{1/2} \quad (7.17)$$

Once the $d_{i,t'_c} \geq wj$, the BS will notice the MSS_i to migrate to a new channel in $Ak(k = [d_{i,t'_c}/w])$ with the message (P'_t, c'_n). For example, the MSS might exceed the boundary of A_j or $d_{i,t'_c} \leq w(j-1)$. Therefore, by using the prediction to prevent the out-of-service effect, the performance of the BWA system can be maintained well. Besides, the overhead of prediction will not be heavy since we only use the routine procedure of channel measurement, which is specified in the IEEE 802.16 standard, to get the information for estimation.

7.5 The Predicted Handover Scheme

Whenever an MSS in roaming between BSs, only two BSs need to be dealing with the handover. Consequently, the MSS should be informed for a crucial message from the only TBS so it can perform a fast handover with the TBS. Based on the above concept, we assume SBS will be allocated for one available channel to MSS_i in area A_{SBS}. $A_{SBS_{10}}$ is the microcell from one of the fragment of a six-piecewise divided macrocell, forming h concentric hexagonal cells with an equal width w. To prevent the out-of-VoIP service effect of MSS_i performing handover, we investigate a PHS developed on a BS-centralized control mechanism to deal with the problem of handover. The SBS controls the location, distance, and direction of the MSS. According to these parameters, the BS will periodically compute the timing of handover (T_{HO}) which is independent of the current moving speed of the MSS. The SBS will always periodically change the T_{HO} on receiving a report response (REP-RSP) message from each MSS. According to the direction of MSS, the SBS will easily select the only TBS. Therefore, SBS will actively coordinate with TBS for the handover of MSS over the backbone.

When the SBS receives all handover-related information of the TBS, it may simultaneously convey to MSS. However, the MSS is required to only transmit the CDMA ranging code at the TBS, as a result, the MSS does not have to wait for the RNG-RSP message from TBS. By using the PHS, the SBS will handle all handover processes of the MSS and allows the MSS to easily use its service and also share a large loading amount of MSS. In the sequence diagram of PHS, steps (f) through (i) are defined in the IEEE 802.16e standard

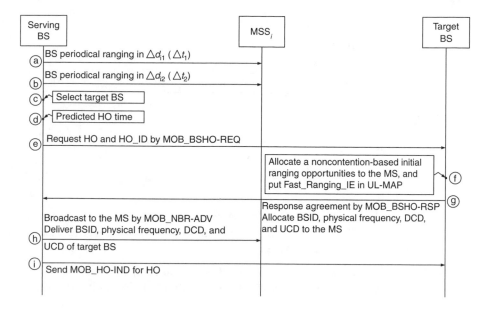

FIGURE 7.4
The sequence diagram of predicted handover scheme.

[16], and we simplify the complex definition. The details of centralized control handover processes of PHS are given in the following nine steps, as shown in Figure 7.4:

- **Step a:** The IEEE 802.16e standard [16] recommends that the BS has to broadcast a report request (REP-REQ) message to all MSSs for channel measurements within 10 s to check whether the MSS is still in the service set. Therefore, the BS can get the SNR value by the replied REP-RSP message from each MSS to estimate the distance periodically. According to Equation 7.17, for location prediction and channel migration, the SBS measures the radio quality of MSSs and then using their SNR, determines the distance of $\Delta d_{i1}(\Delta t_1)$ of MSS_i.

    According to the geographic channel alignment (GCA) framework [7], we can calculate h by

$$h = \left\lfloor \sqrt{\frac{N_C - 3}{3|C_{00}|} + 2} - 1 \right\rfloor \qquad (7.18)$$

    where N_C is the number of channels for usage in a macrocell and the number of channels in C_{00} is represented as $|C_{00}|$. The macrocell's boundary is denoted as d_{cell} and can be obtained by

$$d_{\text{cell}} = d_0 \times 10 \exp \left\{ \left[P_t + G_t + G_r \right. \right.$$
$$- 20 \log \left[\frac{4\pi d_0 \left(\mathcal{F}_H - (3(h-1)^2 |C_{00}| + 1)B \right)}{c} \right.$$
$$\left. \left. - X_\sigma - C_f - C_H - L - \text{SNR}_{r,\min} - N_0 \right] \middle/ 10\rho \right\}$$

$$(7.19)$$

- **Step b:** Similarly, $\Delta d_{i2}(\Delta t_2)$ of MSS_i can be calculated. Thus the movement distance between them is $\Delta d_{i2} - \Delta d_{i1}$, and the time between them is $\Delta t_2 - \Delta t_1 = 10\,\text{s}$.

- **Step c:** The MSS_i drives in the direction of $\overrightarrow{\Delta d_{i1} \Delta d_{i2}}$. Following this direction, the SBS can decide a unique TBS for MSS_i to handover. Details can be found in Ref. 7.

- **Step d:** According to the velocity equation, distance divided by time, we have

$$\frac{\Delta d_{i2} - \Delta d_{i1}}{\Delta t_2 - \Delta t_1} = V_{\text{MSS}_i} \qquad (7.20)$$

The average time of velocity $\text{MSS}_{\text{iv(AV)}}$ will be

$$\overline{V}_{\text{MSS}_i} = \frac{V_{\text{MSS}_{i1}} + V_{\text{MSS}_{i2}} + \cdots + V_{\text{MSS}_{ix}}}{x}, \qquad x \in 1, 2, 3, \ldots \quad (7.21)$$

By using Equation 7.5 and SBS, we can predict T_{HO} of MSS_i, denoted as

$$\frac{d_{\text{cell}} - \Delta d_{i2}}{\overline{V}_{\text{MSS}_i}} = \text{MSS}_i T_{\text{HO}} \qquad (7.22)$$

- **Step e:** When MSS_i approaches $\Delta d_{\text{HO}} \geq d_{\text{cell}} \div h \times (h-1)$, SBS requests precoordination to TBS for handover and HO_ID by MOB_BSHO-REQ message, which includes channel quality information channel identifier (CQICH_ID) assigned to the MSS_i as identification. The Δd_{HO} is the boundary h of a macrocell.

- **Steps f and g:** If the resource of TBS is available, TBS will allocate a noncontention-based initial ranging opportunity to the MSS_i. Synchronously, TBS puts fast ranging information element (fast_ranging_IE message) in UL-MAP and responds agreement to SBS by handover (HO) respond (MOB_BSHO-RSP) message, which allocates BSID, physical frequency, DCD, and UCD of TBS. For this precoordination, TBS will hold the request service for 10 s.

- **Step h:** SBS will prepare precoordination handover message of boundary MSS every 10 s. When MSS requests HO, SBS broadcasts responded agreement message from TBS to MSS$_i$ by the neighbor advertisement (MOB_NBR-ADV) message, which includes BSID, physical frequency, DCD, and UCD of TBS.

- **Step i:** When MSS$_i$ receives the TBS message, MSS$_i$ will immediately send an HO indication (MOB_HO-IND) message, which includes BSID, physical frequency, DCD, UCD, and fast_ranging_IE message of TBS for HO with TBS. If the TBS is available for MSS$_i$, MSS$_i$ can quickly enter TBS without preceding CDMA-based initial raging by a nonzero value of fast_ranging_IE parameter at MSS$_i T_{HO}$ in approaching d_{cell}. The MSS$_i$ will migrate to a new channel in A_{TBS}.

In view of the centralized control handover processes of PHS as shown in Figure 7.4, we can get the service DT for PHS (DT_{PHS}) during the handover process by

$$DT_{PHS} = T_{SBS \rightarrow MSS} + T_{SYN} \tag{7.23}$$

7.6 Simulation Handover Model and Results

We use the QualNet 3.9.5 developer command-line simulator and design new embedded handover module of PHS, L_0, and L_2 to simulate average service DT and handover, dropping probabilities during handover process. In our simulation model, there are seven BSs each of them dominating a hexagon cell and six hexagons are around one hexagon. The diameter of each hexagon is 1000 m long. The range of operating spectrum is from 2.40 to 2.46 GHz and is divided by a fixed bandwidth (BW) 10 MHz into several independent channels. The simulation model is operating in TDD mode. The fast fourier transform (FFT) (N_{FFT}) size is 1024. The sampling frequency (F_s) can be calculated by $F_s = (n * BW * 8000)/8000$ as 11.42 MHz. The subcarrier spacing (Δf) can be calculated by $\Delta f = F_s/N_{FFT}$ as 11.16 kHz. T_u, the useful symbol time can be calculated by $T_u = 1/\Delta f$ as 89.64 μs. Guard time $T_g = T_u/8$ as 11.2 μs. Each orthogonal frequency-discussion multiple access (OFDMA) symbol time (T_s) is evaluated by $T_s = T_u + T_g$ as 100.84 μs. The downlink and uplink (DL/UL) ratio is 3:2. The number of subchannels is 30.

The OFDMA frame length is 5 ms and is also the minimum one-time transmission unit. Therefore, any message transfer must follow frame by frame and the time of one-way transmission cannot be less than 5 ms. The initial BS's transmission power of the BS is 300 mW. The simulation model-specific parameters of the IEEE 802.16e MAC protocol we used are shown in Table 7.2.

TABLE 7.2

Parameters Used in Disruption Time

Parameter	Value
Spectrum (GHz) (for 7 BSs)	2.40–2.46
The distance between two BSs (m)	1000
Bandwidth (MHz) (BW)	10
FFT size (N_{FFT})	1024
DL/UL ratio	3:2
OFDMA frame length (ms)	5
Sampling frequency (MHz)	11.42
Subcarrier spacing (kHz)	11.16
Useful symbol time (μs)	89.64
Guard time (μs)	11.2
OFDMA symbol time (μs)	100.84
Number of subchannels	30
Number of OFDMA symbol per frame	49
CW_{min} (opportunities)	32
CW_{max} (opportunities)	1024
CW request oppurtunity per frame (OFDMA symbols)	12
Maximum number of CW request retry	10
Ranging opportunity per frame (OFDMA symbols)	12
Maximum number of ranging retry	10
Average time of contention ranging (ms) T_{CR}	120
Average time of reauthorization (ms) T_{RA}	175
Rendezvous time (ms) T_R	50
Average time of reregistration (ms) T_{RR}(2frames)	35
Average time of Internet delay (ms) T_{ID}	50
Average time of RNG-REQ (ms) T_{RNG}	25
Average time of frame synchronize (ms) T_{SYN}	5
MSS$\leftrightarrows$SBS (ms) (1frame*2way)	10
SBS$\leftrightarrows$nBS (ms) (1frame*2way)	10
SBS$\leftrightarrows$TBS (ms) (1frame*2way)	10
MSS$\leftarrow$SBS (ms) (1frame)	5

The simulation environment is built by one serving BS with 40 MSSs and 6 nBSs concurrently within a 1500×1500 m square. All MSSs are randomly developed around the BS. All MSSs execute the ranging request process by adopting QPSK 1/2 encoding rate.

Figure 7.5 illustrates the average service DT during handover process under a given number of MSSs with a fixed speed of 100 h/km. The DT parameters of the IEEE 802.16e standard we used are shown in Table 7.2. The service DT of L_0 and L_1 are much larger than that of L_2 and PHS, because of the long reauthorization and reregistration process. As shown in the figure, the minimum disruption time of PHS reaches 11 ms when M reaches 40. The reason why PHS outperforms L_0, L_1, and L_2 is that PHS considers a BS centralized control mechanism to predict T_{HO} and to deal with the problem of handover beforehand. As a result, PHS will accommodate more MSSs and thus get less DT as the number of MSSs increases. On the contrary, L_0, L_1, and L_2 only reach their minimum disruption time at 335, 300, and 70 ms due to long process.

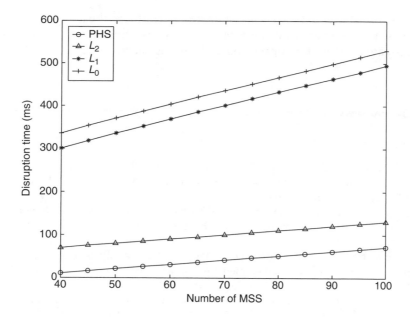

FIGURE 7.5
Disruption time versus M.

In addition, PHS still outperforms L_2 as $M \leq 100$ (largest number of MSSs) due to the effect of the appropriate centralized management in SBS. This scheme will avoid useless handover processes and transform useless messages to MSSs, which will get lower disruption times. The saving time is achieved by precoordination of reauthorization and reregistration with the TBS and remitting the time of ranging with the TBS. Therefore, the service DT of PHS is less than any other handover schemes.

7.7 Conclusion

In this chapter, we investigated a measured signal-aware mechanism for BS, which periodically monitors on moving MSSs and prepares CDMA ranging code for handover use beforehand. Simulation results show that PHS decreases the average service DT of the WiMAX as well as lowers handover failure probability of MSSs efficiently even in highly competitive circumstance. Through the derived system model expression, we present the PHS scheme to improve the lower T_{DT} by close to 40 ms without changing the standard IEEE 802.16e standard. Specifically, in our proposed solutions the MAC protocol at both the BS and MSS do not need to be modified and are readily disposable over the existing network infrastructure.

Simulations show that the PHS system model confirms the analytical results. Moreover, by considering the mobility of MSSs, the PHS scheme can be investigated further for supporting QoS among macrocells.

References

1. I.F. Akyildiz and W. Wang, The predictive user mobility profile framework for wireless multimedia networks, *IEEE/ACM Trans. Network.* Vol. 12, No. 6, pp. 1021–1035, December 2004.
2. J.B. Andersen, T.S. Rappaport, and S. Yoshida, Propagation measurements and models for wireless communications channels, *IEEE Commun. Mag.*, Vol. 33, No. 1, pp. 42–49, January 1995.
3. N. Banerjee, K. Basu, and S. Das, Handoff delay analysis and measurement in SIP-based mobility management in wireless networks, in *Proc. Int. Parallel Distrib. Process. Symp.*, pp. 224–231, April 2003.
4. J.J. Caffery and G.L. Stüber, Overview of radiolocation in CDMA cellular systems, *IEEE Commun. Mag.*, Vol. 36, No. 4, pp. 38–45, April 1998.
5. C.-J. Chang, T.-T. Su, and Y.-Y. Chiang, Analysis of A cutoff priority cellular radio system with finite queueing and reneging/dropping, *IEEE/ACM Trans. Network.* Vol. 2, pp. 166–175, April 1994.
6. J. Chen and W.-K. Tan, Predictive dynamic channel allocation scheme for improving power saving and mobility in BWA networks, *ACM/Springer Mobile Networks and Applications (MONET)*, 2006.
7. J. Chen, C.-C. Wang, and J.-D. Lee, Geographic channel assignment framework for broadband wireless access networks, *IEICE Trans. Commun.*, Vol. E89-B, No. 11, pp. 3160–3163, November 2006.
8. K.-H. Chiang and N. Shenoy, A 2-D random-walk mobility model for location-management studies in wireless networks, *IEEE Trans. Veh. Technol.*, Vol. 53, No. 2, pp. 413–424, March 2004.
9. S. Choi, G.-H. Hwang, T. Kwon, A.-R. Lim, and D.-H. Cho, Fast handover scheme for real-time downlink services in IEEE 802.16e BWA system, in *Proc. IEEE VTC 2005-Spring*, Vol. 3, pp. 2028–2032, Stockholm, Sweden, May 2005.
10. T.-S. Chu and L.J. Greenstein, A quantification of link budget differences between the cellular and PCS bands, *IEEE Trans. Veh. Technol.*, Vol. 48, No. 1, pp. 60–65, January 1999.
11. C. Eklund, R.B. Marks, K. L. Standwood, and S. Wang, IEEE Standard 802.16: A technical overview of the wirelessman air interface for broadband wireless access, *IEEE Commun. Mag.*, Vol. 40, No. 6, pp. 98–107, June 2002.
12. V. Erceg, L.J. Greenstein, S.Y. Tjandra, S.R. Parkoff, A. Gupta, B. Kulic, A.A. Julius, and R. Bianchi, An empirically based path loss model for wireless channels in suburban environments, *IEEE J. Select. Areas Commun.*, Vol. 17, No. 7, pp. 1205–1211, July 1999.
13. V. Erceg, L.J. Greenstein, S.Y. Tjandra, S.R. Parkoff, A. Gupta, B. Kulic, A.A. Julius, and R. Bianchi, A model for the multipath delay profile of fixed wireless channels, *IEEE J. Select. Areas Commun.*, Vol. 17, No. 3, pp. 399–410, March 1999.

14. C. Evci and B. Fino, Spectrum management, pricing, and efficiency control in broadband wireless communications, *Proc. IEEE*, Vol. 89, No. 1, pp. 105–115, January 2001.
15. R.A. Guerin, Queueing-blocking system with two arrival streams and guard channels, *IEEE Trans. Commun.*, Vol. 36, pp. 153–163, February 1988.
16. IEEE 802.16 Working Group, *IEEE Standard for Local and Metropolitan Area Networks—Part 16: Air Interface for Fixed and Mobile Broadband Wireless Access Systems, Amendment 2: Physical and Medium Access Control Layers for Combined Fixed and Mobile Operation in Licensed Bands and Corrigendum 1*, IEEE Std. 802.16e–2005, February 2006.
17. A. Kavak, M. Torlak, W.J. Vogel, and G. Xu, Vector channels for smart AntennasXMeasurements, statistical modeling, and directional properties in outdoor environments, *IEEE Trans. Microwave Theory Tech.*, Vol. 48, No. 6, pp. 930–937, June 2000.
18. W.K. Lai and J.C. Chiu, Improving handoff performance in wireless overlay networks by switching between two-layer IPv6 and one-layer IPv6 addressing, *IEEE J. Select. Areas Commun.*, Vol. 23, No. 11, pp. 2129–2137, November 2005.
19. R. Laroia, S. Uppala, and L. Junyi, Designing a mobile broadband wireless access network, *IEEE Signal Process. Mag.*, Vol. 21, No. 5, pp. 20–28, September 2004.
20. P.N. Pathirana, A.V. Savkin, and S. Jha, Location estimation and trajectory prediction for cellular networks with mobile base stations, *IEEE Trans. Veh. Technol.*, Vol. 53, No. 6, pp. 1903–1913, November 2004.
21. M. Pätzold and N. Youssef, Modelling and simulation of direction-selective and frequency-selective mobile radio channels, *Int. J. Electron. Commun.*, Vol. 55, No. 6, pp. 433–442, December 2001.
22. G. Plitsis, Coverage prediction of new elements of systems beyond 3G: The IEEE 802.16 system as a case study, in *Proc. IEEE VTC 2003-Fall*, Vol. 4, pp. 2292–2296, Orlando, Florida, October 2003.
23. T. S. Rappaport, *Wireless Communications: Principles and Practice*, Prentice Hall PTR, Upper Saddle River, New Jersey, 1996.
24. S. Tekinay and B. Jabbari, A measurement-based prioritization scheme for handovers in mobile cellular networks, *IEEE J. Select. Areas Commun.*, Vol. 10, pp. 1343–1350, October 1992.
25. A.E. Xhafa and O.K. Tonguz, Dynamic priority queueing of handover calls in wireless networks: An analytical framework, *IEEE J. Select. Areas Commun.*, Vol. 22, No. 45, pp. 904–916, June 2004.

8

802.16 Mesh Networking

Petar Djukic and Shahrokh Valaee*

CONTENTS

8.1 Introduction

Wireless mesh networks interconnect access points (APs) spread out over a large geographical area. Wireless terminals (WTs) connect to the APs on

---

* This work was sponsored in part by LG Electronics Corporation.

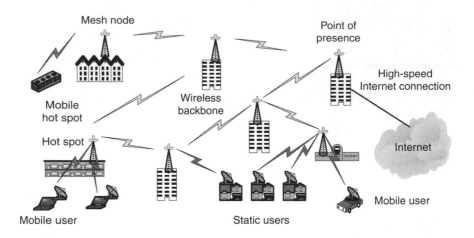

FIGURE 8.1
A mesh network has a number of static backbone nodes that carry traffic for users in the network. Each WT connects to an AP at the edge of the network, and this AP sends WT's traffic over the backbone to the point-of-presence, which is connected to the Internet. Since there is only one high-speed Internet connection for many APs, the network has a low operational cost.

their first hop. Then, their traffic is carried by the wireless mesh to the point-of-presence (POP) where it can go to the Internet (Figure 8.1). The POP is the only node in the network connected to the Internet and can act as a base station (mesh coordinator). In urban areas, mesh networks interconnect wireless hot spots. Mesh networks decrease the cost of running the hot spots since they only require a single POP broadband connection for the whole network. For example, using a mesh network to interconnect 133 existing hot spots in the Toronto downtown area would decrease the total cost of running the hot spots by 70% [1]. Mesh networks can also be used to provide the wireless last mile in rural areas where it is impractical to provide wired connectivity due to sparseness of customers. This is the idea behind rooftop networks [2], where each house has a mesh node connecting it to neighboring houses while providing wireless access to the devices in the house.

Current mesh networks use 802.11 technology to interconnect the mesh backbone [3,4]. However, 802.11 technology is a decade old and was not designed for mesh networks. In particular, 802.11 lacks the extensions to provide quality-of-service (QoS) in multihop wireless environments [5]. The 802.11 protocol also lacks security extensions needed to provide WTs with privacy and security across the mesh backbone. These problems are addressed by the 802.16 mesh technology [6]. IEEE 802.16 uses time division multiple access (TDMA) technology to provide QoS and encryption for security and privacy. This chapter reviews 802.16 mesh technology and proposes solutions needed in the network layer to take advantage of 802.16 mesh extensions.

IEEE 802.16 mesh uses TDMA technology to provide link-level QoS in the network. In TDMA, QoS required by WTs is negotiated in terms of end-to-end bandwidth reserved for each WT on links connecting it to the POP. QoS is enforced at each link with scheduled access to the wireless channel. Link bandwidth is allocated over frames with a fixed number of slots and a scheduler assigns slots to links. During each slot, a number of links that do not conflict with each other may transmit simultaneously. Two links conflict with each other if transmissions by one link prevent packet reception at the other. The bandwidth of each link is given by the number of slots assigned to it in the frame and the modulation used in the slots.

The 802.16 mesh protocol specifies two scheduling protocols for assignment of link bandwidths: centralized and decentralized scheduling protocols. The centralized scheduling protocol is used by the base station (mesh coordinator) to establish network-wide schedules. In contrast, the decentralized scheduling protocol is used to negotiate pairwise bandwidth assignments between mesh routers. The centralized scheduling protocol can be used to establish network-wide end-to-end QoS; however, the decentralized scheduling protocol is not expected to establish end-to-end QoS.

In 802.16, links between routers are managed with logical connections. Logical connections are established between mesh routers within the wireless range of each other and remain valid as long as the network operates. However, a connection may be inactive if it is not assigned any TDMA slots. Using a connection-oriented protocol is appropriate for mesh networks since mesh routers are usually static with respect to each other. The connection-oriented nature of 802.16 protocol significantly improves the efficiency of the mesh. For example, the protocol uses a combination of an 8-bit network ID 16-bit mesh ID, and an 8-bit link ID to associate transmissions with links, compared to 48-bit Ethernet address pairs used by 802.11.

Since 802.16 is a connection-oriented protocol, the network stack used on 802.11 mesh nodes is not applicable for 802.16 networks for several reasons. First, 802.16 mesh networks do not have layer-2 broadcast capabilities and use a convergence sublayer (CS) to multiplex Internet protocol (IP) packets to connections. Therefore ARP [7] is not needed. Second, when a medium access control (MAC) layer scheduling algorithm changes the state of a connection, the routing protocol used on the node should be notified of the change so that routes can be adjusted accordingly. A change in link status may propagate routing changes, which affects QoS. It is therefore necessary to design a network layer that is aware of the TDMA nature of 802.16 networks. Third, since 802.16 mesh networks are intended for infrastructure-based mesh networks, the 802.16 routers are static and always on, meaning that the connection lifetime is in the order of the network lifetime. The scale of the connection lifetime makes it possible to establish hop-by-hop security in the mesh backbone, by keeping a private key in sync on both sides of the connection. In 802.16, private keys are distributed and managed with a key management protocol initiated by the base station.

This chapter reviews the networking aspects of 802.16 mesh networks with a focus on exposing scheduling, routing, and security problems in the protocol. We describe the current state-of-the-art research addressing the problems, and we propose our solutions to some of the problems left open in the standard. Section 8.2 presents an overview of TDMA technology used in 802.16 mesh networks and the scheduling algorithms required to manage TDMA slots. We review the current research analyzing the scheduling algorithms provided in the standard. We also review the research proposing scheduling algorithms required by the standard but left open to the implementation. Section 8.3 presents an overview of the network layer architecture in 802.16 mesh networks, including routing and addressing issues introduced by TDMA technology. The 802.16 standard specifies that the IP layer should be connected to the 802.16 hardware with a CS; however, the implementation details of the CS are left out. We specify a CS that takes advantage of QoS inherently available in 802.16 mesh protocol and integrates it with IP DiffServ architecture [8]. Section 8.4 presents an overview of the security architecture in 802.16 mesh networks and the research exposing the security flaws in the standard. We present our security additions, which enhance end-to-end security in the network layer.

8.2   802.16 Time Division Multiple Access

In this section, we describe the 802.16 mesh TDMA MAC technology and the research problems posed by this technology. First, we describe the orthogonal frequency division multiplexing (OFDM) technology at the physical layer, which provides equal-duration time division multiplexing (TDM) slots required for TDMA. We then summarize research activities toward alternative technologies that can provide TDM timing for 802.16 MAC. Second, we describe how TDM slots are grouped into frames and how transmissions are scheduled with logical TDMA channels. The 802.16 standard specifies scheduling algorithms for the logical channels used for mesh control messages. We outline the current research into the performance of those algorithms. The scheduling algorithms for data channels are left entirely to the implementation of the standard. We summarize the research proposing TDMA scheduling algorithms for 802.16 networks. We conclude the section with a description of how nodes are assigned their initial TDMA bandwidth when they enter the network, which resolves a practical problem often ignored in research.

8.2.1   802.16 Physical Layer

IEEE 802.16 is a TDMA-based MAC protocol built on a TDM physical layer. In TDM physical layers, the time is divided into time slots of equal length, and during each time slot, a block of bytes is transmitted. IEEE 802.16 uses OFDM to implement the TDM physical layer. OFDM transforms blocks of bits

TABLE 8.1

Comparison of 802.11a and 802.16 Raw Data Rates

		Raw Bitrate (Megabits/Second)	
Modulation	Bits/Symbol	10 MHz Bandwidth	20 MHz Bandwidth
BPSK-1/2	96	3.84	7.68
QPSK-1/2	192	7.68	15.36
QPSK-3/4	288	11.52	23.04
16QAM-1/2	384	15.36	30.72
16QAM-3/4	576	23.04	46.08
64QAM-2/3	768	30.72	61.44
64QAM-3/4	864	34.56	69.12

into constant-duration symbols carried on multiple, orthogonal carriers. The bandwidth of the final signal is the frequency range occupied by the carriers. Bandwidth used by 802.16 OFDM may be allocated in the license-exempt 5 GHz frequency band or in other, licensed, frequency bands.

The number of raw bits carried by each OFDM symbol depends on the modulation, coding, and the bandwidth occupied by OFDM during transmissions (Table 8.1). Modulation and coding determine how many bits are carried by each orthogonal carrier, while OFDM bandwidth dictates the duration of the symbols. In 802.16, there are two possible OFDM bandwidths: 20 MHz with 12.5 μs symbol duration and 10 MHz with 25 μs symbol duration. Since the OFDM symbol duration for 10 MHz bandwidth is twice as long as the OFDM symbol duration for 20 MHz bandwidth, the raw bitrate at 10 MHz bandwidth is half of the raw bitrate at 20 MHz bandwidth.

In situations where 802.16 hardware is not available, but QoS in the mesh is still required, it is possible to use 802.16 TDMA technology with properly controlled 802.11 hardware. In Ref. 9, we have shown that it is possible to embed 802.16 MAC packets into 802.11a [10] broadcast packets with insignificant overhead. TDM is achieved by fixing the 802.11 back-off mechanism to one slot before every transmission. The back-off time limits can be changed on 802.11 hardware supporting QoS. We have shown in Ref. 9 that the system with embedded 802.16 packets has a performance comparable to the performance of true 802.16 systems. In Ref. 11, the authors show that drivers of a specific 802.11 network card can be modified so that true TDM is achieved over 802.11 hardware. This is different from the approach in Ref. 9 since it requires that the details of the operation of the underlying 802.11 hardware be available at the MAC layer.

8.2.2 TDMA Framing and Transmission Timing

OFDM symbols are grouped into TDMA frames of equal length and the frames are repeated over time (Figure 8.2). OFDM symbols in each frame are divided

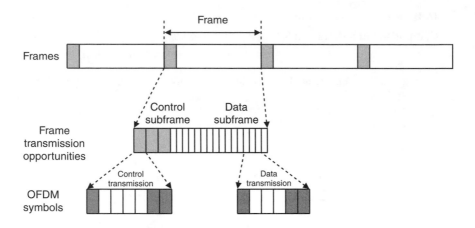

FIGURE 8.2
802.16 Time Division Multiple Access (TDMA). 802.16 uses OFDM to achieve TDMA. OFDM symbols are grouped into frames of fixed duration. Frames are logically divided into the control subframe and the data subframe. In the control subframe, transmission opportunities are 7 OFDM symbols long. The length of transmission opportunities in the data subframe depends on the number of OFDM symbols in the frame. In this example, the length of transmission opportunities in the data frame is 2 OFDM symbols.

into two subframes. The first part of the frame is the control subframe, used to transmit 802.16 control packets. The second part of the frame is the data subframe, used to transmit data packets. There are two types of control subframes and the whole network alternates between them. The first type of control subframe is the scheduling subframe in which nodes transmit scheduling messages. The second type of control subframe is the network configuration subframe in which nodes broadcast network configuration packets containing topology information, network provisioning information, and network management messages. The network configuration subframes occur periodically with the period indicated with the parameter SchedulingFrames. SchedulingFrames is a network parameter transmitted in the network configuration subframe.

The management of OFDM symbols is simplified by grouping them into transmission opportunities. In the control subframe, the symbols are grouped into transmission opportunities with a fixed length of 7 OFDM symbols. Four of the symbols are used to transmit information at the lowest bitrate, while the other three are used as guard symbols (Figure 8.2). There are a total of MSH-CTRL-LEN transmission opportunities in each control subframe, where MSH-CTRL-LEN is a network parameter transmitted in the network configuration subframe. In the data subframe, the symbols are grouped into transmission opportunities whose length depends on the number of OFDM symbols in the frame. For example, in Figure 8.2, the data transmission is 3 transmission opportunities long, corresponding to 6 OFDM symbols. The size of data transmission opportunities is found by dividing the number of data symbols in the frame by 256 and taking the integer part of the result:

$$\text{DataTxOppSize} = \left\lfloor \frac{N_f - 7 \times \text{MSH-CTRL-LEN}}{256} \right\rfloor \quad (8.1)$$

where N_f is the number of OFDM symbols in the frame and $7 \times \text{MSH-CTRL-LEN}$ is the number of OFDM symbols in the control subframe. The reason for limiting the number of transmission opportunities in the data subframe to 256 is that fields referring to transmission opportunities in 802.116 scheduling packets are 8-bits long.

Transmission opportunities are assigned to logical channels. There are three types of logical channels: basic, broadcast, and data. The basic channel is used for ranging and network entry packets, the broadcast channel is used to transmit mesh control packets, and the data channels are used for data packets and some 802.16 control packets. The basic channel is allocated in the control subframe. Some of slots for the broadcast channels are in the control subframe and some are in the data subframe. All data channel slots are located in the data subframe. The basic channel and the data channels are unicast since only one node is supposed to process transmissions from the channel, while messages in the broadcast channel are intended for all first-hop neighbors of a node.

The channels are closely related to the types of packets transmitted in them; we summarize the relationship between the mesh control packet types and channel types in Table 8.2. The basic channel is used by nodes entering the network to transmit the network entry MSH-NENT packets. Broadcast channels are used to transmit MSH-NCFG, network configuration messages, and MSH-CSCF, MSH-CSCH, and MSH-DSCH scheduling messages. There are three types of broadcast channels depending on how transmission opportunities in the channel are shared. There are two reliable broadcast channels that use coordinated transmissions to prevent collisions. The first uses distributed election-based scheduling for MSH-NCFG and MSH-DSCH messages. The second uses tree-based scheduling for MSH-CSCH and MSH-CSCF messages. Optionally, MSH-DSCH messages can also be transmitted in the unused

TABLE 8.2

802.16 Mesh Control Packets

Packet Type	Channel Type	Scheduling	Purpose
MSH-NENT	Basic	Best effort	Network entry
MSH-NCFG	Broadcast	Distributed election	Network configuration
MSH-CSCF	Broadcast	Tree scheduling	Centralized scheduling configuration
MSH-CSCH	Broadcast	Tree scheduling	Centralized schedule distribution
MSH-DSCH	Broadcast	Distributed election, Best effort	Decentralized scheduling negotiation

data slots of the data subframe in an additional unreliable broadcast channel. We elaborate on the purpose of each of the control packets further in the rest of this section.

8.2.3 Transmission Scheduling in the Logical Channels

Each channel type has its own method to assign transmission opportunities to the nodes sharing the channel. The assignment method is usually specified in terms of a scheduling protocol and a scheduling algorithm. The rest of the section elaborates on the scheduling protocols and algorithms used for each channel type.

8.2.3.1   *The Basic Channel*

The basic channel is used by nodes entering the network to transmit the network entry MSH-NENT packets. The basic channel transmission opportunities are allocated in the first control transmission opportunity of every network configuration frame (Figure 8.3). The basic channel is a best effort channel, so it does not guarantee collision-free transmissions. Nodes transmitting in the basic channel use a 1 s timer to retransmit unacknowledged packets. The 802.16 mesh standard does not specify a back-off mechanism for the basic channel, even though such a mechanism may be useful in case the channel is busy. The standard assumes that the mesh nodes are static and that they are always on. However, if 802.16 mesh nodes are used in scenarios where the mesh nodes are mobile, or frequently off, a back-off mechanism can be added to the channel without any changes to the 802.16 standard.

8.2.3.2   *Distributed Election Scheduling Broadcast Channels*

Network configuration messages (MSH-NCFG) and coordinated distributed scheduling messages (MSH-DSCH) use broadcast channels with distributed

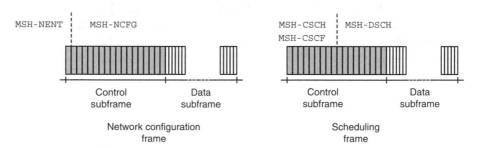

FIGURE 8.3

Placement of logical channels in the frame. The first transmission opportunity of every network configuration frame is reserved for MSH-NENT messages (basic channel); the other transmission opportunities in the frame are reserved for MSH-NCFG messages (broadcast channel). All transmission opportunities in the control subframe of the scheduling frames are reserved for the broadcast channel. However, the last MSH-DCSH-NUM control transmission opportunities are reserved for MSH-DSCH messages.

election scheduling. In distributed election scheduling, each transmitter sharing the channel broadcasts the range of opportunities it considers for transmissions. The transmitters whose ranges of transmission opportunities overlap with their two-hop neighbor's ranges perform a distributed election procedure for each transmission opportunity. The election algorithm guarantees that each transmission opportunity has only one winner, so that the transmissions in the channel are collision-free. We first describe how transmission opportunities are related to actual OFDM symbols and then describe how 802.16 distributes the ranges of transmission opportunities and elects winners of each transmission opportunity.

The network configuration broadcast channel is located in the control subframe of every network configuration frame (Figure 8.3). The transmission opportunities can be viewed on their own axis if we ignore all of the OFDM symbols not used by the channel (Figure 8.4). Given the index of a transmission opportunity in the channel, CurrentTxOpp, the frame in which the transmission should take place can be found by dividing CurrentTxOpp by the number of network configuration transmission opportunities in each frame, MSH-CTRL-LEN − 1, and then multiplying by the number of frames between successive network configuration frames. The index of the starting OFDM symbol for the transmission can be found by subtracting the number of transmission opportunities before the start of the frame and then multiplying by 7 to account for the length of each transmission opportunity.

The distributed scheduling broadcast channel is located in the last $7 \times$ MSH-DSCH-NUM OFDM symbols of the control scheduling subframes, after the centralized scheduling messages (Figure 8.3). MSH-DSCH-NUM is a network parameter indicating the number of transmission opportunities in the control subframe allocated to distributed scheduling messages. As with

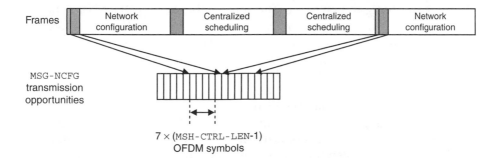

FIGURE 8.4
The MSH-NCFG transmission opportunities are mapped from OFDM symbols in the control subframe to the logical transmission opportunities on the MSH-NCFG axis. On the MSH-NCFG axis, the transmissions are indexed as a continuous set of integers starting with 0. In this example, MSH-CTRL-LEN = 6 and SchedulingFrames = 2.

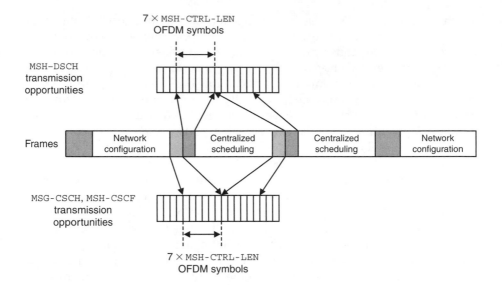

FIGURE 8.5

The `MSH-CSCH`, `MSH-CSCF` and `MSH-DSCH` transmission opportunities are mapped to two different transmission opportunity axes. On the `MSH-CSCH`, `MSH-CSCF` axis, the transmission opportunities are assigned with the tree scheduling algorithm. On the `MSH-DSCH` axis, the transmission opportunities are assigned with the distributed election algorithm. In this example, `MSH-CTRL-LEN = 6`, `SchedulingFrames = 2`.

the `MSH-NCFG` channel, the transmission opportunities in the distributed scheduling channel can be viewed on their own axis (Figure 8.5). Given the index of a transmission opportunity in the channel, `CurrentTxOpp`, the frame in which the transmission should take place can be found by taking a modulus with respect to `SchedulingFrames` and adding 1 to account for the network configuration frame. The index of the starting OFDM symbol can be found by subtracting the number of transmission opportunities before the start of the frame and then adding the number of OFDM symbols used for the centralized scheduling channel.

In both the network configuration and distributed scheduling broadcast channels, transmission opportunities are assigned with the use of a distributed election algorithm. The distributed election algorithm specified in the 802.16 mesh standard works in two parts. First, the nodes exchange the range of transmission opportunities they consider for transmission. Second, the nodes contending for the same transmission opportunity perform an election to decide who should transmit during the conflicting transmission opportunity. The election procedure uses a combination of the conflicting transmission opportunity index and each of the conflicting node identifiers to create a unique, pseudorandom, 16-bit hash value. The node with the highest 16-bit hash value for the transmission opportunity wins the election.

For the election procedure to be deterministic, all nodes must have the same view of which transmission opportunities are in dispute. The information about available transmission opportunities is distributed in a two-hop neighborhood of every node. Each node transmits a range of transmission opportunities it considers for transmission in terms of lower and upper bounds. The nodes also rebroadcast the ranges of transmission opportunities of their one-hop neighbors, so that the transmission opportunities are known throughout the two-hop neighborhood of the nodes. Since the size of control packets is limited, the 802.16 mesh standard specifies that the range of contended transmission opportunities should be compressed into a 3-bit maximum value, Mx, and a 5-bit hold-off exponent, He. Given the encoding for the range, the minimum number of transmission opportunities before the next transmission by a node is calculated with

$$\text{MinNextXmtTime} = 2^{\text{He}+4} + \text{Mx} \times 2^{\text{He}} \tag{8.2}$$

and the maximum number of transmission opportunities is calculated with

$$\text{MaxNextXmtTime} = 2^{\text{He}+4} + (\text{Mx}+1) \times 2^{\text{He}} \tag{8.3}$$

Potential transmission conflicts can be found since all nodes broadcast their Mx and He values as well as rebroadcast all of their immediate neighbor's Mx and He values. Given the ranges of potential transmissions for their two-hop neighborhood, nodes can check if their choice of next transmission time in the channel conflicts with any transmissions with

$$2^{\text{He}_i+4} + \text{Mx}_i \times 2^{\text{He}_i} \leq \text{NextOpportunity} \leq 2^{\text{He}_i+4} + (\text{Mx}_i+1) \times 2^{\text{He}_i} \tag{8.4}$$

where Mx_i and He_i are associated with the two-hop neighbor i, and NextOpportunity is the opportunity the node is considering for its next transmission. To avoid collisions with neighbors whose Mx_i and He_i are not known, the nodes assume that those neighbors transmit all the time.

Performance of the distributed election scheme is analyzed in Ref. 12. In that work, the authors derive an analytical expression for the average time required to access the distributed election channel. The authors use a partial 802.16 mesh simulator to measure the distributed election access times and compare the simulations to theoretical results. The simulations show that the theoretical model is fairly accurate. The paper also points out that the expected time to access the channel depends on the He value, so flows requiring quicker access to the channel should use smaller values of He.

8.2.3.3 Tree-Based Scheduling Broadcast Channels

Centralized scheduling MSH-CSCH and MSH-CSCF messages are transmitted in a tree-based scheduling broadcast channel. Scheduling of transmissions in this channel is performed by following a breadth-first traversal of a globally known tree topology. The global tree topology is first distributed with MSH-CSCF messages, which carry the entire routing tree the messages are multicast on. If the topology changes, further MSH-CSCF messages notify the nodes of the changes. As nodes receive MSH-CSCF messages, they learn the multicast routing tree, as well as which node in the topology is currently broadcasting the MSH-CSCF message, so they can calculate the transmission opportunity in which they should transmit. In the case of MSH-CSCH messages, the nodes know the topology prior to any transmission of messages, so they can also calculate their next transmission opportunity the same way they would for MSH-CSCF messages.

8.2.3.4   Best Effort Broadcast Channel

The best effort broadcast channel is used for transmission of distributed scheduling MSH-DSCH messages. This channel consists of unused transmission opportunities in the data channel. The 802.16 standard does not set any rules on how this channel should be accessed.

8.2.3.5   Transmission Scheduling in the Data Channels

In 802.16 mesh protocol, there are two types of data channels: the centralized scheduling data channel and the distributed scheduling data channel. The difference between the two data channels is in how their transmission opportunities are assigned. In the centralized scheduling data channel, the transmission opportunities are assigned with the centralized scheduling protocol, which relies on the base station to assign connection bandwidths and distribute them to all nodes. The nodes use the knowledge of the bandwidth assignments to independently calculate the global transmission schedule. In the distributed scheduling channel, the transmission opportunities are distributed with the decentralized scheduling protocol, which uses pairwise negotiation of connection bandwidths to achieve conflict-free schedules using only local information.

The centralized scheduling data channel uses the first MSH-CSCH-DATA-FRACTION transmission opportunities of the data subframe. These transmission opportunities are assigned with the centralized scheduling protocol. In the centralized scheduling protocol, the nodes request bandwidth from the base station by sending MSH-CSCH messages to their parents in the scheduling tree (Figure 8.6). Once all the requests reach the base station, the base station uses them to calculate the bandwidth for each connection in the network and multicasts the connection bandwidth assignments using a new wave of MSH-CSCH messages. The connections whose centralized scheduling bandwidth is positive, form a tree coinciding with the scheduling tree for MSH-CSCH and MSH-CSCF messages. If by changing the connection bandwidths, the base station also changes the routing tree for the network, it multicasts routing changes with MSH-CSCF messages before it multicasts MSH-CSCH messages. MSH-CSCH messages coming from the base station

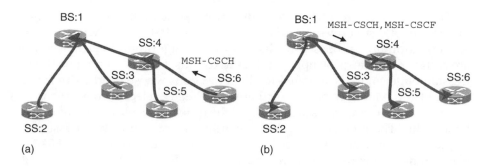

FIGURE 8.6

802.16 centralized scheduling. Mesh nodes send requests to the base station with MSH-CSCH messages moving up the tree. The base station uses the information from the received MSH-CSCH messages together with its knowledge of network topology to calculate a global TDMA schedule for the data subframe. The base station then updates the tree topology with the MSH-CSCF messages and transmits new bandwidth assignments with the MSH-CSCH messages. The nodes use the link bandwidths to find the transmission schedule. (a) Up-tree scheduling messages. (b) Down-tree scheduling messages.

contain connection bandwidths for every connection in the network, so each node can run an independent scheduling algorithm to arrive at a global transmission schedule. The new schedule takes place in the first frame after the last node in the tree receives its MSH-CSCH message.

The 802.16 standard does not specify how connections should be assigned their bandwidth; however, it does propose an algorithm that the nodes can use to determine a transmission schedule for the entire network given an assignment of connection bandwidths. The scheduling algorithm proposed in Ref. 6 uses a breadth-first traversal of the routing tree to assign transmission opportunities for all connections in the network. The first-visited connection, in the traversal of the tree, is assigned transmission opportunities at the beginning of the data subframe. The number of transmission opportunities needed to satisfy the bandwidth allocation B for the connection are found with

$$\texttt{Duration} = \left\lceil \frac{BT_f/b + N_g}{\texttt{DataTxOppSize}} \right\rceil \tag{8.5}$$

where $\lceil \cdot \rceil$ denotes the ceiling of a real number, b the highest number of bits that can be transmitted in each OFDM symbol on the connection, DataTxOppSize the number of OFDM symbols in each transmission opportunity, N_g the number of guard OFDM symbols (two or three), and T_f the frame duration in seconds. The connection traversed next is assigned next available transmission opportunities and so on, until all connections are assigned the number of transmission opportunities corresponding to their bandwidth. If the total length of the schedule exceeds MSH-CSCH-DATA-FRACTION transmission opportunities, all connection bandwidths are scaled equally until the schedule is at most MSH-CSCH-DATA-FRACTION transmission opportunities long.

The scheduling algorithm in Ref. 6 does not take advantage of spatial reuse in the network, so it does not efficiently use the wireless bandwidth. A different algorithm is proposed in Ref. 13. In that algorithm, the connections are assigned transmission opportunities in rounds. In each round, one transmission opportunity is allocated to all connections whose bandwidth is not satisfied and which are not conflicting with already-selected connections in the round. The connections are chosen in the order of decreasing unallocated bandwidth. The problem with this scheduling algorithm is that it assumes connections can transmit more than once in a frame. However, in 802.16, every transmission needs a guard time of two or three TDMA slots, meaning that at the highest modulation each transmission has an overhead of 216 or 324 bytes. The overhead decreases the value of the algorithm in Ref. 13. We propose an algorithm that can be used to find a global schedule in Ref. 14. Our algorithm limits the number of connection transmissions to one per frame. The algorithm uses the Bellman–Ford algorithm on the conflict graph for the network to find starting transmission opportunities for each connection. In Ref. 14, we also give a set of simple linear inequality constraints that guarantee that an allocation of connection bandwidths results in a feasible schedule. The base station can use the linear constraints to ensure that bandwidth assignments result in TDMA schedules without the need to scale down link bandwidths.

The transmission opportunities after the first MSH-CSCH-DATA-FRACTION opportunities in the data channels are reserved with distributed scheduling. In distributed scheduling, nodes negotiate the distribution of transmission opportunities in a pairwise fashion. First, a node wishing to change the transmission opportunity allocation for one of its connections sends a request for transmission opportunities to its neighbors in an MSH-DSCH packet. One or more of the neighbors correspond with a range of available transmission opportunities. The node chooses a subrange of the available transmission opportunities and confirms that it will use them with a third MSH-DSCH packet. The 802.16 standard does not specify the algorithms that can be used to calculate which slots should be requested or released during the distributed scheduling. We provide a distributed scheduling algorithm in Ref. 15 that can be adapted for this purpose. In our algorithm, we use a distributed Bellman–Ford algorithm to iteratively find the TDMA schedule from connection demands. The advantage of our algorithm is that the algorithm requires only a partial knowledge of the network topology, available from 802.16 neighbor tables, to determine a conflict-free TDMA schedule.

The centralized and distributed scheduling give rise to two different QoS levels in the mesh network. Connections established with the centralized scheduling protocol have a guaranteed bandwidth, granted by the base station and known throughout the network. The hop-by-hop bandwidth guarantee in the centralized scheduling routing tree allows end-to-end QoS guarantees for the traffic flows traversing the tree. However connections established with the decentralized scheduling protocol have a transient behavior and a bandwidth dependent on the grants from the node's neighbors. The uncertainty in connection bandwidth translates to the best effort QoS to end-to-end flows using the connection scheduled with the distributed scheduling protocol.

An important question in the design of 802.16 mesh networks is the number of transmission opportunities in the data channel that should be allocated for guaranteed traffic. Clearly, MSH-CSCH-DATA-FRACTION should be minimized so that as much bandwidth as possible be available for best effort traffic and enough bandwidth can be allocated for the services requiring guaranteed bandwidth. We minimize the number of slots needed to schedule links in the centralized scheduling part of the data frame in Ref. 14. The optimization finds the smallest value of MSH-CSCH-DATA-FRACTION required to support the requested link bandwidths, subject to the limit on TDMA propagation delay in the network. TDMA propagation delay occurs when an outgoing link on a mesh node is scheduled to transmit before an incoming link in the path of a packet [14].

8.2.4 Network Entry and Synchronization

Since 802.16 is a collision-free, TDMA-based protocol, careful network entry is required to ensure that new nodes do not disrupt TDMA transmissions that are already scheduled. The network entry procedure in the 802.16 mesh standard specifies the stages of entry for the new node and logical channels the nodes can access during the entry procedure (Figure 8.7). Initially, a node wishing to enter the network (candidate node) synchronizes itself to the frame boundary by listening to MSH-NCFG packets from the nodes already in the network. When the candidate node is synchronized to the frame boundary, it can use the basic channel to start the network entry procedure.

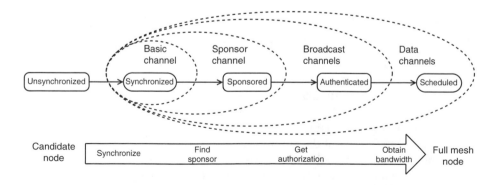

FIGURE 8.7
States of 802.16 network entry. Initially, the candidate node is only allowed to use the basic channel. After finding a sponsor, it uses the sponsor channel to authenticate with the base station. Once it is authenticated, the candidate node becomes a full-fledged mesh node and it is allowed to use the broadcast channel to get a bandwidth assignment in the data channel.

The candidate node selects the first mesh router that it receives two consecutive MSH-NCFG packets from as its sponsor node. The role of the sponsor node is to be an intermediary between the candidate node and the rest of the mesh network, by allocating a part of its reserved data channel as a special sponsor channel for the candidate node. The candidate node transmits an MSH-NENT packet to the potential sponsor node indicating that it wishes to enter the network. The sponsor node checks the credentials of the candidate node, received in the MSH-NENT packet, and if the node decides to become a sponsor, it transmits a sponsoring confirmation in one of its MSH-NCFG packets. The sponsoring confirmation includes the range of data channel transmission opportunities that the sponsoring node assigns to the candidate node during its network entry. The candidate node uses the sponsor channel to authenticate itself with the base station. (We give the details of the authentication process in Section 8.4.) After the candidate node is authenticated, it can start using the broadcast channel to transmit its MSH-NCFG and scheduling messages, so it closes the sponsoring channel with the final MSH-NENT packet in the basic channel.

Network synchronization is achieved with MSH-NCFG packets. MSH-NCFG packets are broadcast regularly and each packet includes a summary of the two-hop neighborhood for the node. As a part of the neighbor information, nodes transmit their propagation delay estimates for each neighbor. Each MSH-NCFG packet also includes the number of hops from the sending node to the base station. Given the timing information in the MSH-NCFG packets, the nodes can synchronize with the base station. Each node synchronizes to MSH-NCFG packets from the neighbor closest to the base station, and can use the propagation delay estimate from the synchronizing node to itself to adjust its timing to match the network timing.

8.3   802.16 Mesh Networking

We have shown in the previous section that the 802.16 mesh standard has cross-layer design features, such as centralized scheduling, that cross the boundary between the MAC layer and the IP layer on the mesh nodes. These types of cross-layer features can be used to enhance the QoS in the mesh if they are taken advantage of. In this section, we show how to design the addressing in the network layer so that the network takes full advantage of QoS available with 802.16 MAC and yet the 802.16 mesh routers can be simple, in line with the mesh network application scenarios outlined in Section 8.1. We also design the CS, which allows the network layer to access 802.16 QoS features.

8.3.1   802.16 MAC Connections

The 802.16 mesh standard uses a combination of a 16-bit mesh identifier (ID) and a 16-bit connection identifier (CID) to identify the source and destination of every transmission. Mesh ID is a unique mesh node identifier obtained during the authentication process and is assigned by the base station. The CID is calculated dynamically and it depends on the type of channel the transmission is in. In the data channel, the CID refers to a logical data connection between two neighbors. In this case, the CID is a combination of an 8-bit link ID and an 8-bit QoS description for the connection. The 8-bit link ID identifies the receiver of the connection, relative to the sender of the packet. In the basic channel and the broadcast channel, the CID is a combination of an 8-bit network ID and 0xFF (meaning any link ID). In the basic channel, the receiver of the transmission is identified with its 16-bit mesh ID in the MSH-NENT packet, deviating from the way receivers are identified in data channel unicast connections.

Data connections are established between neighbors with a sender-initiated negotiation. First, the sender initiates a link creation with a request in one of its MSH-NCFG packets. The request includes a hashed message authentication code (HMAC) for the request message, obtained by applying a network-wide secret key obtained during the authentication process [15]. The receiver checks the request and if it can recalculate the HMAC, it responds with a positive response in one of its MSH-NCFG packets. Finally, the initiator sends an 8-bit link ID it will use to refer to the connection in subsequent data transmissions. In subsequent data transmissions, the 8-bit link ID is extracted from the CID so that a node can identify its packets.

A unicast data connection between two mesh nodes can be in one of four states after it is created. First, it could have no bandwidth allocated to it. In this case, the connection cannot be used to transfer data, so it is in the DOWN state (Figure 8.8). Second, it could have bandwidth allocated to it with the centralized scheduling protocol. In this case, it is in the UP-CSCH state. Third, it could have bandwidth allocated to it with the decentralized scheduling protocol.

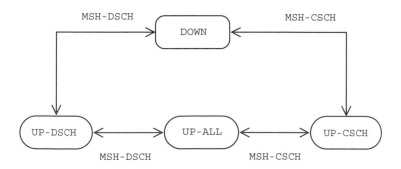

FIGURE 8.8
State transitions for 802.16 connections. Connection change state after receiving one of the scheduling messages. Centralized scheduling messages bring a connection to and from the UP-CSCH state. Decentralized scheduling messages bring a connection to and from the UP-DSCH state. If a connection is already scheduled by centralized or decentralized scheduling, it may be assigned bandwidth with the other scheduling protocol and come into the UP-ALL state.

In this case, it is in the UP-DSCH state. Finally, the connection could have bandwidth allocated to it with both centralized and decentralized scheduling protocol. In this case, it is in the UP-ALL state.

Change of state for a connection causes routing changes in the network. For example, if a connection goes from the DOWN state to one of the three up states, this adds a new neighbor in the network layer. Similarly, if a connection goes from one of the three up states to the DOWN state, this change removes a neighbor in the network layer. Neighbor connections do not change in the MAC layer, since MAC layer neighbors communicate in the 802.16 broadcast channel.

The combination of the mesh ID and the CID identifies each connection globally, so we refer to the 32-bit value of the combination as the global connection ID (GCID). Using fixed connection identifiers for addressing is appropriate for mesh networks since links between neighbors are static. It is also more efficient than a pair of 48-bit MAC addresses used to identify source and destination pairs in 802.11 networks.

8.3.2   Mesh Network Addressing

The 802.16 mesh network protocol specifies how addressing is accomplished in the MAC layer. We now propose a network layer addressing scheme that keeps mesh routers simple to implement. We partition the network into access networks and the mesh backbone. There is an access network at every mesh node, allowing the WTs to connect to the mesh in the network layer (Figure 8.9). To keep the address space for the whole network small, each

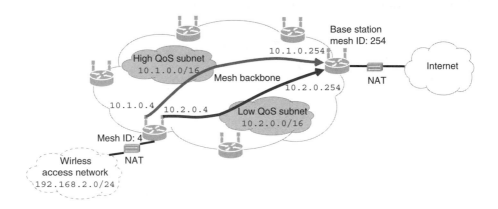

FIGURE 8.9
Subnets in a 802.16 mesh network. The backbone network uses two subnets, one with high QoS (centralized scheduling) and one with low QoS (distributed scheduling). The wireless terminals connect to mesh nodes over the wireless access subnet. Mesh nodes use network address translation to forward packets from the access network to the backbone network. The base station also uses NAT when forwarding packets from the mesh backbone to the Internet.

access network uses the same block of private IP addresses in the range 192.168.2.0/24 [16]. The mesh nodes use network address translation (NAT) [17] to allow WTs to access the mesh backbone.

In the backbone, mesh routers use a range of private addresses different from the range in the access networks. We assume that mesh routers are assigned addresses in the range 10.0.0.0/8 with the assignment of IP addresses closely matching the assignment of mesh IDs. We set the last 16 bits of the IP address to the mesh ID and keep bits 8–15 of the addresses for subnet classification. For example, address 10.1.0.4 corresponds to the address of mesh router 4 on subnet 1. Since the base station acts as a POP for the network, it also provides NAT services for the packets traversing the mesh backbone.

We assign each mesh router to multiple subnets to simplify how QoS is enforced in the network (Figure 8.9). We use a subnet providing a low QoS and a subnet providing a high QoS; however, a number of subnets with different QoS may be larger if necessary. We use the source marking model of QoS [8], where WTs mark the level of service they require in the type of service (TOS) field of their outgoing IP packets [8,18]. Mesh routers examine the TOS field of packets coming from their access network and, depending on the QoS specified, either forward the packet over the high QoS subnet or the low QoS subnet. This way, *all* per packet QoS decisions are made at the edge of the mesh backbone and the forwarding engine on each mesh router is simplified. The QoS classification of packets is done by the routing module, before the convergence sublayer, which we describe next.

8.3.3 QoS-Aware Convergence Sublayer

The 802.16 standard specifies that the IP layer should be connected to the 802.16 MAC layer with a CS, which classifies packets to connections, based on their CID. The standard omits the details of how the CS should operate. In this section, we propose a CS that uses a combination of logical interfaces and QoS subnets to take advantage of 802.16 QoS in the network layer.

The CS is designed to work together with the 802.16 scheduler, since scheduling changes may affect QoS. For example, in centralized scheduling, a large number of connections may change their state at the same time. A large number of simultaneous changes in the entire topology would cause a wave of updates in a dynamic routing protocol such as OSPF [19]. While the routing tables are converging, data packets may bounce around the network, causing large delays. However, in decentralized scheduling, only a few connections change status at any given time. This is a normal operation of the MAC layer expected from dynamic routing protocols and consistent with the QoS provided by decentralized scheduling.

We address the QoS issues caused by changes in connection state with a QoS-aware convergence CS. The CS resides in the operating system of mesh routers (Figure 8.10). Our CS is a combination of logical interfaces provided to the network layer (IP) and the way the interfaces classify packets. There is one

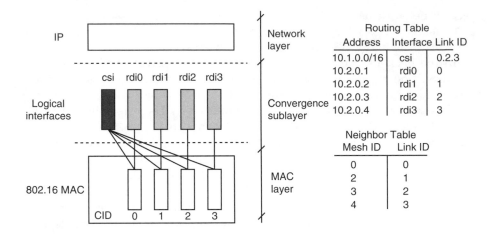

FIGURE 8.10
Routing and neighbor tables in a 802.16 router. The routing table resides in the network layer and it associates network addresses to interfaces. For the DCI, the routing table associates the entire high QoS subnet range with the interface. For normal interfaces, the routing table associates individual neighbor IP addresses with the interface. The neighbor link table resides in the MAC layer and it matches mesh IDs to links. Mesh IDs are obtained from the last 16 bits of the IP address.

logical interface for all packets traversing the centralized scheduling routing tree (centralized scheduling interface [CSI]) and multiple logical interfaces for each data connection on the router (data interfaces [DIs]).

The idea behind having a special interface for all connections using centralized scheduling is to hide the routing taking place in the MAC layer from the network layer. The routing table assigns the high-quality QoS subnet to the CSI, so that packets traversing the high QoS subnet are forwarded through the CSI. When a packet comes to CSI, the interface finds out the mesh ID of destination from the IP destination address and forwards the packet on the logical connection that is the next hop in the centralized scheduling tree. The CSI interface presents the network layer with static routes along the centralized scheduling routing tree, even though the actual routes may change if the link schedule changes. In effect, the CSI interface performs bridging for the high QoS traffic in the network.

Data interfaces have a one-to-one correspondance with the 802.16 connections. When an IP packet is forwarded to a DI, it is transmitted to the mesh router linked with the corresponding logical connection, so DIs are assigned peer addresses corresponding to their 802.16 peers. Since, some of the connections may have their bandwidth assigned with the centralized scheduling and others may have bandwidth assigned with decentralized scheduling, there are no guarantees on how the DI bandwidth is assigned. However, if connections are restricted from being in the UP-ALL state, no connection will be assigned both centralized and decentralized bandwidths. We also add a

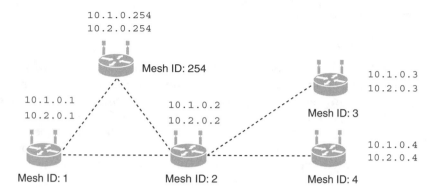

FIGURE 8.11
Network scenario for Figure 8.10.

restriction that if a connection is assigned bandwidth with the centralized scheduling, its corresponding DI is in the DOWN state. This way all packets traversing the low QoS subnet will be transmitted on connections scheduled with the distributed scheduling protocol.

Figures 8.10 and 8.11 show how the logical interfaces and routing work together in practice. Figure 8.10 shows the routing table and the neighborhood table for node 2 in Figure 8.11. Each mesh node belongs to both the high and the low QoS subnets. The routing table associates the entire high QoS subnet with the CSI and it also associates each of the connections established with the node's neighbors individually as peer-to-peer links in the low QoS subnet. When an IP packet arrives to an interface, the mesh ID of the final destination for the packet can be found by extracting the last 16 bits of the destination address. In the case of CSI, the next-hop link ID is found from the routing table of mesh IDs that CSI obtains from centralized scheduling messages. In the case of a DI, the link ID is obtained from the interface number associated with DI. The CID for the MAC layer data transmission can be obtained by combining the mesh ID of the node 2 with the link ID of the logical connection to the next-hop router.

The needs to notify the routing protocols when one of the connections changes status from DOWN to UP-DSCH, or if the connection changes state from UP-DSCH to DOWN. We do not specify the exact notification method since it is operating system specific. For example, in the Linux operating system, it is sufficient to change the status of the logical MAC interface, which automatically updates the forwarding tables in the kernel [20]. The CS also needs to notify the network layer if all of the connections whose bandwidth was scheduled with the centralized scheduling protocol change state from UP-CSCH to DOWN. In this case, the CSI interface becomes unavailable since the node is disconnected from the centralized scheduling routing tree.

8.4 Network Security

The 802.16 MAC protocol specifies security procedures used to authenticate new nodes and exchange and maintain private encryption keys. The private encryption keys are used to encrypt traffic to first-hop neighbors or to the base station. We first review the authentication process during which network-wide shared secrets are distributed to mesh routers entering the mesh. We then review how private keys are exchanged between peer nodes so that 802.16 peers can encrypt data packets. Finally, we propose an end-to-end security scheme to simplify security in the mesh.

8.4.1 Network Authentication

Before nodes can use the network, they authenticate themselves with the base station. The authentication of new mesh nodes is performed with the privacy key management (PKM) protocol [6]. The PKM protocol is also used to distribute and maintain private keys used for traffic encryption.

During network entry, the new node (candidate node) first finds a sponsor node, which provides a portion of its own bandwidth as the sponsor channel. The candidate node uses the sponsor channel to authenticate with the base station (Figure 8.12). The candidate node sends a PKM-REQ packet to the authentication server, which may reside on the base station. Since the candidate node may not be directly connected to the base station, and the authentication server, the sponsor node tunnels the candidate's PKM-REQ packet to the base station with UDP. The PKM-REQ message carries a X.509 certificate [21] belonging to the candidate node. The X.509 certificate is used to verify the authenticity of the candidate node and it also contains the candidate's public RSA key. If the candidate node is verified, the authentication

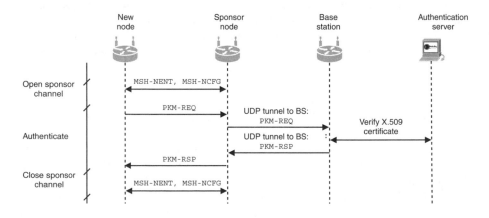

FIGURE 8.12
Mesh node authorization.

server sends back a `PKM-RSP` message containing an operator-shared secret, the list of security associations, identified with their security associated identifiers (SAIDs), and authorization keys (AKs), one for each SAID, all encrypted with the candidate's public RSA key.

The operator-shared secret is used to validate nodes during the link establishment process; it is used to calculate the HMACs for the link establishment messages. Security associations are used to manage encryption information for connections and to assign AKs to connections. AKs are used to derive key encryption keys (KEKs) for subsequent PKM communications, as well as to validate PKM communications within the security association with HMACs. The security associations have a limited lifetime, so the PKM protocol requires the nodes to periodically reauthorize and get new AKs.

The base station always sends a primary security association and it may optionally send other static security associations. The primary security association is used for communications with the base station. Static security associations are used for data traffic. If the base station does not send any static traffic security association, the nodes use the primary security association. With our QoS scheme, there could be two security associations. The first one can be used for connections in the high QoS subnet and the other can be used for connections in the low QoS subnet.

Network authorization is vulnerable to "man-in-the-middle" attack. Specifically, since the X.509 certificate sent by the entering node contains the public key of the new node, a malicious node can masquarade itself as the authentication node and give the false security settings [22]. The reason this type of attack is possible is that there is no mutual trust between the new node and the authentication node. The new node must assume that the authentication response is indeed from the authorization node. A modification to the PKM protocol that removes these types of attacks from 802.16 mesh networks is proposed in Ref. 22. In this version of the PKM protocol, the authentication server sends its certification to the candidate node, allowing the new node to authenticate the authenticator and thus establish mutual trust.

8.4.2 Backbone Hop-by-Hop Security

Data communications in 802.16 mesh networks are protected with hop-by-hop encryption of packets. Data can be encrypted with 56-bit DES or with the AES CCM algorithm. In each case, the encryption is accomplished with a shared, private, traffic encryption key (TEK) for the connection. TEKs are generated independently on the nodes with a pseudorandom algorithm specified in Ref. 23. The PKM protocol specifies the mechanism for TEK exchange between nodes.

TEKs are exchanged between MAC layer neighbors. A node initiates the exchange by sending a key request to its neighbor. The key request message contains the sender's X.509 certificate and a hash value calculated with the AK that the sender obtained from the base station during authorization.

If the neighbor node can authorize the request, by verifying the hash with its AK, it sends back a TEK encrypted with a KEK. The authentication of the packet verifies that both nodes are still authenticated with the base station. If one of the nodes is using an expired AK, the peer node will find out from the incorrect HMAC value for the packet.

The TEK is encrypted with one of three algorithms: 1024-bit RSA, 3-DES, or 128-bit AES. The key encryption method is assigned through the security association the connection is in. If 1024-bit RSA encryption is used for TEK encryption, the node sending the TEK uses the RSA public key that the requesting node sends in its X.509 certificate as the KEK. If 3-DES encryption is used for TEK encryption, the node sending the TEK uses the AK it obtained during the authorization from the base station to generate a private key. The private key is generated by first padding the AK with 0x63 repeated 64 times, taking the SHA-1 hash of the result and truncating it to 128 bits. If 128-bit AES encryption is used for the TEK, the KEK is obtained in the same way as for 3-DES encryption.

8.4.3   User End-to-End Security

IEEE 802.16 provides a mechanism to encrypt traffic traversing data connections at each hop. However, manufacturing mesh routers that can perform encryption at high speeds, available at the physical layer, may be costly. In this section, we propose an end-to-end encryption in the network layer that takes the encryption out of the mesh backbone. In our scheme, encryption is handled at the edge of the network. WTs establish a VPN tunnel with a VPN server outside of the mesh backbone, so no encryption is required by mesh routers.

We add a VPN server after the POP, but before the traffic goes on the Internet (Figure 8.13). The server is on a special unprotected subnet 10.0.0.0/16. The WTs negotiate IPSec tunnels with the VPN server, and after the VPN tunnel is established, the WTs get an IP address on the protected subnet 10.254.0.0/16. The VPN tunnel may be in any mode, e.g., encryption

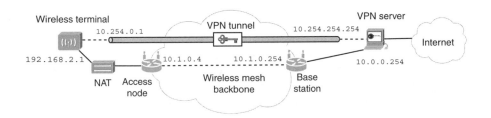

FIGURE 8.13
End-to-end VPN tunneling. WTs connect to the VPN server, which is on an unprotected network (10.0.0.0/16). Once a WT establishes a VPN tunnel with the server, it is assigned an IP address on the protected (10.254.0.0/16) subnet.

of the payload or both IP headers and the payload. This means that both the client WT and the authentication server should support IPSec NAT traversal (NAT-T), which allows the use of IPSec over NAT [24,25]. This is not a problem since NAT-T is a part of modern operating systems [26]. The QoS is oblivious to IPSec since the type of service field is copied from the header of the inner, plain text IP packet to the header of the outer, encrypted, packet [27].

Although this end-to-end encryption scheme protects WT traffic, it does not protect the 802.16 mesh management traffic. This means that 802.16 nodes should still use the primary security association encryption to communicate with the base station. However, since this presents a small amount of traffic, implementing it in practice may not be hard.

8.5   Conclusion

We have reviewed the 802.16 mesh protocol. This protocol uses TDMA to provide QoS in the mesh backbone. QoS is available in the MAC layer, so we have introduced a subnetting scheme in the network layer that takes advantage of the QoS. The subnetting allows us to move all QoS decisions to the edge of the network. We have also proposed a CS that glues together the subnetting in the network layer to different classes of service available in 802.16 mesh networks.

IEEE 802.16 provides for hop-by-hop encryption of data traffic; however, this may be costly to implement in practice. We have proposed an end-to-end security scheme that takes encryption away from the mesh backbone to the clients using the network. This should significantly simplify the implementation of mesh routers.

References

1. M. Chee, *The Business Case for Wireless Mesh Networks*, www.nortelnetworks. com, December 2003.
2. B. A. Chambers, *The Grid Roofnet: A Rooftop ad hoc Wireless Network*, M. Eng., Massachusetts Institute of Technology, 2002.
3. J. Camp, J. Robinson, C. Steger, and E. Knightly, *Measurement Driven Deployment of a Two-Tier Urban Mesh Access Network*, Rice University, Technical Report TREE0505, December 2005.
4. Nortel Networks, *Wireless Mesh Network—Extending the Reach of Wireless LAN, Securely and Cost-Effectively*, http://www.nortelnetworks.com/solutions/ wlan/, November, 2003.
5. S. Xu and T. Saadawi, Does the IEEE 802.11 MAC protocol work well in multihop wireless ad hoc networks, *IEEE Communications Magazine*, vol. 39, no. 6, pp. 130–137, June 2001.

6. *IEEE Standard for local and metropolitan Area Networks Part 16: Air Interface for Fixed Broadband Wireless Access Systems*, 2004.

7. D. C. Plummer, *Ethernet Address Resolution Protocol: Or Converting Network Protocol Addresses to 48.bit Ethernet Address for Transmission on Ethernet Hardware*, RFC 826 (Standard), http://www.ietf.org/rfc/rfc826.txt, November 1982.

8. Y. Bernet, P. Ford, R. Yavatkar, F. Baker, L. Zhang, M. Speer, R. Braden, B. Davie, J. Wroclawski, and E. Felstaine, *A Framework for Integrated Services Operation over Diffserv Networks*, RFC 2998 (informational), http://www.ietf.org/rfc/rfc2998.txt, November 2000.

9. P. Djukic and S. Valaee, 802.16 MCF for 802.11a based mesh networks: A case for standards re-use, in *23rd Queen's Biennial Symposium on Communications*, 2006.

10. *IEEE Standard for Local and Metropolitan Area Networks Part 11: Wireless LAN Medium Access Control (MAC) and Physical Layer (PHY) Specifications High-Speed Physical Layer in the 5 GHz Band*, 1999.

11. M. Neufeld, J. Fifield, C. Doerr, A. Sheth, and D. Grunwald, SoftMAC—flexible wireless research platform, *HotNets*, 2005.

12. M. Cao, W. Ma, Q. Zhang, X. Wang, and W. Zhu, Modeling and performance analysis of the distributed scheduler in IEEE 802.16 mesh mode, *MobiHoc*, 2005, pp. 78–89.

13. H.-Y. Wei, S. Ganguly, R. Izmailov, and Z. Haas, Interference-aware IEEE 802.16 WiMax mesh networks, *VTC Spring'05*, 2005.

14. P. Djukic and S. Valaee, *Quality-of-Service Provisioning in Multi-Service TDMA Mesh Networks*, http://www.comm.utoronto.ca/~djukic/Publications/publications.html, University of Toronto, WIRLab Technical Report, August 2006.

15. H. Krawczyk, M. Bellare, and R. Canetti, *HMAC: Keyed-Hashing for Message Authentication*, RFC 2104 (informational), http://www.ietf.org/rfc/rfc2104.txt, February 1997.

16. Y. Rekhter, B. Moskowitz, D. Karrenberg, and G. de Groot, *Address Allocation for Private Internets*, RFC 1597 (informational), obsoleted by RFC 1918. http://www.ietf.org/rfc/rfc1597.txt, March 1994.

17. P. Srisuresh and K. Egevang, *Traditional IP Network Address translator (Traditional NAT)*, RFC 3022 (informational), http://www.ietf.org/rfc/rfc3022.txt, January 2001.

18. J. Postel, *Internet Protocol*, RFC 791 (Standard), updated by RFC 1349. http://www.ietf.org/rfc/rfc791.txt, September 1981.

19. J. Moy, *OSPF Version 2*, RFC 1247 (draft standard), obsoleted by RFC 1583, updated by RFC 1349. http://www.ietf.org/rfc/rfc1247.txt, July 1991.

20. *Linux kernel*, http://www.kernel.org/, 2006.

21. R. Housley, W. Polk, W. Ford, and D. Solo, *Internet X.509 Public Key Infrastructure Certificate and Certificate Revocation List (CRL) Profile*, RFC 3280 (proposed standard), updated by RFCs 4325, 4630. http://www.ietf.org/rfc/rfc3280.txt, April 2002.

22. S. Wattanachai, *Security Architecture of the IEEE 802.16 Standard for Mesh Networks*, M.Sc., Royal Institute of Technology, Stockholm University, April 2006.

23. D. Eastlake III, S. Crocker, and J. Schiller, *Randomness Recommendations for Security*, RFC 1750 (informational), obsoleted by RFC 4086. http://www.ietf.org/rfc/rfc1750.txt, December 1994.

24. T. Kivinen, B. Swander, A. Huttunen, and V. Volpe, *Negotiation of NAT-Traversal in the IKE*, RFC 3947 (proposed standard), http://www.ietf.org/rfc/rfc3947.txt, January 2005.

25. A. Huttunen, B. Swander, V. Volpe, L. DiBurro, and M. Stenberg, *UDP Encapsulation of IPsec ESP Packets*, RFC 3948 (proposed standard), http://www.ietf.org/rfc/rfc3948.txt, January 2005.

26. *IPsec NAT Traversal Overview*, ser. The Cable Guy Column. Microsoft TechNet, http://www.microsoft.com/technet/, August 2002.

27. D. Maughan, M. Schertler, M. Schneider, and J. Turner, *Internet Security Association and Key Management Protocol (ISAKMP)*, RFC 2408 (proposed standard), obsoleted by RFC 4306. http://www.ietf.org/rfc/rfc2408.txt, November 1998.

9

WiMAX Testing

Rana Ejaz Ahmed

CONTENTS

9.1   Introduction

Testing and certification of telecommunication products have been a challenging task owing to the rigorous and complex nature of the testing process and the related infrastructure involved. Telecommunication product manufacturers often spend a large portion of their time and budget on testing activities, as they are very critical to the overall success and marketability of the product. The testing time of wireless products can easily account for 70% of the cost of the product as engineers test for certification, government compliance, and electromagnetic compatibility and electromagnetic interference mitigation [13]. WiMAX technology is a standard-based form of wireless broadband in which products from different vendors are intended to be interoperable, thus boosting competition and driving down prices through high-volume product production. WiMAX product testing and certification are more complicated due to complexity in several factors, including radio, protocol, and interoperability testing issues.

The telecommunications industry tests its products for performance, interoperability, conformance, integration, stress, volume, etc. Conformance testing is the act of determining to what extent a single implementation conforms to the individual requirements of its base standard [1]. In the case of WiMAX, the conformance testing may include unit testing, mandatory regulatory/compliance type testing, and testing against underlying standards. The role of conformance testing is to increase the confidence that the product conforms to its specifications, and to minimize the risk of malfunctioning when the product is put into place. Interoperability testing verifies if the end-to-end functionality between (at least) two implementations of communicating systems is as required by those base systems' standards. It is to be noted that interoperability testing is not a substitute for conformance testing. Both conformance and interoperability testing are needed, as one can argue that two implementations following the same wrong specifications could be still interoperable.

One of the main elements of WiMAX technology is the interoperability of WiMAX products, certified by the WiMAX Forum [2], resulting in mass volume and confidence for the service providers to buy equipment from more than one company and that such integration works together. The WiMAX Forum defines and conducts conformance and interoperability testing to ensure that different vendor systems work seamlessly with one another. Those that pass conformance and interoperability testing achieve "WiMAX Forum Certified" designation and can display this mark on their products and marketing materials. The certification program demonstrates a certain measure of compliance and interoperability. However, since it is unrealistic and impractical to test products for every single aspect of the specifications, certification does not provide full guarantee. It gives a reasonable and acceptable degree of confidence [3].

This chapter provides details about the different types of testing and certification needed for WiMAX products. It also describes some major test equipment used for WiMAX product testing.

9.2   Conformance Testing

Conformance testing is the verification that a unit under test (UUT) (i.e., a WiMAX product/implementation, system, or a subsystem) meets the formal requirements of the protocol standard (derived from protocol implementation conformance specifications, PICS). The family of standards related to IEEE 802.16 (IEEE standard for local and metropolitan area networks—Part 16: Air interface for fixed broadband wireless access systems) [9] or ETSI HiperMAN standards apply for conformance testing.

The conformance testing at the vendor site may include several other types of testing, such as functional (unit) testing, performance testing, stress testing, etc. In this section, we focus primarily on the testing conforming to standards. The conformance testing involves the following phases:

- Regulatory type testing
- Functionality and performance testing for UUT

Regulatory type testing verifies whether the UUT meets the regulations of the country where the product is going to be deployed. This type of testing may include tests for compliance for RF frequency spectrum usage, RF emission monitoring and control (EMC), specific absorption rate (SAR), and other safety regulations used in that country. Different frequency bands are allocated for WiMAX in different parts of the world, and a country may also impose a limit on the maximum power transmitted at the subscriber station (SS) or the base station (BS). Moreover, different countries may have different limitations on the modulation schemes and channelization used. A regulatory body in each country decides the approval procedures. For example, in the European Union, the R&TTE directive describes safety (including SAR), RF, and EMC standards; in the United States, the related FCC regulations are 47CFR (Parts 2, 15, 27, 90, etc.).

The functionality and performance testing are needed to verify whether the UUT meets the standards specifications, more specifically, with respect to the following factors:

- Radio conformance
- Technology family and modulation type (e.g., OFDM/OFDMA, QPSK/16QAM/64QAM)
- Access method (e.g., time division duplexing (TDD)/frequency division duplexing (FDD)); Regulators typically mandate the use of either TDD or FDD
- System capacity, bitrate

Several test equipment from commercial vendors are available to help in the WiMAX testing. The test equipment includes spectrum analyzer, vector signal analyzer (VSA), WiMAX protocol conformance tester, radio conformance tester, WiMAX protocol sniffer, and WiMAX performance and stress test equipment.

9.3 Interoperability and Certification Testing

9.3.1 WiMAX Certification Overview

One of the key elements of WiMAX technology is the interoperability of WiMAX equipment certified by the WiMAX Forum. Without a certification program, it would be very difficult to ensure that the equipment interoperate without going through independent testing. In any standards-based technology, equipment vendors try their best to develop products that comply with the standard. However, different interpretations of standards can lead to lack of interoperability among their products [4].

The WiMAX certification program, launched by WiMAX Forum, is designed to address the conformance and interoperability issues by encouraging cooperation among vendors through "Plugfest" events, where they can informally verify interoperability, and through the formal and official certification testing. The certification program was launched in mid-2005. The certification process includes the following two types of tests that focus on the physical (PHY) and medium access control (MAC) layers:

- Conformance testing to ensure that products correctly implement the specifications defined by IEEE 802.16 and ETSI HiperMAN standards. The vendor is required to complete the PICS questionnaire to specify which features have been implemented in the product for conformance testing. Based on the results of conformance testing, the vendors may choose to modify their hardware or firmware and formally resubmit their products for conformance testing.

- Interoperability testing to verify that the products from different vendors work correctly within the same network. At least three vendors have to submit products within the same certification profile (defined by RF spectrum band, channelization, and duplexing mode used) to start interoperability testing.

The certification testing is conducted at independent labs recommended by the WiMAX Forum. CETECOM lab in Spain was the first lab to perform certification testing, while recently, the Telecommunications Technology Association (TTA) in South Korea was added as the second certification lab. The WiMAX Forum plans to announce additional labs in the future [4]. After a successful certification testing, vendors receive a WiMAX Forum certificate and a test report.

The WiMAX certification process is summarized in Figure 9.1. The roles for the main players in the certification program are summarized in Table 9.1.

It is to be noted that not all WiMAX products will (or are expected to) interoperate with each other. For example, a subscriber unit operating at 3.5 GHz band will not be able to establish a connection with a 5.8 GHz BS; nevertheless, both products are based on the same standards (IEEE 802.16 and ETSI Hiper-MAN) and meet the same requirements. The WiMAX Forum has defined the following two types of *profiles* to address the need of different classes of the products that use the same technology: system profiles and certification profiles. Working with profiles make interoperability more effective and focused.

9.3.2 System Profiles

System profiles set a basic level of common requirements that all WiMAX systems have to meet. To date, only one system profile has been defined and it is based on the IEEE 802.16-2004. A second system profile is currently being defined and will be based on the IEEE 802.16e-2005 and scalable frequency division multiple access. The first system profile is optimized for fixed and nomadic access; the second profile is for portable and mobile access, but also supports fixed and nomadic access [5].

The system profiles define the key mandatory and optional features that are tested in WiMAX products. The features listed as optional in the standards may be tested as mandatory by the certification program; however, the certification does not include any new feature that is not included in the standards. For example, the fixed WiMAX profile based on IEEE 802.16-2004 only allows testing on equipment using point to multipoint operations up to 11 GHz, while IEEE 802.16-2004 equipment can operate up to 66 GHz. The

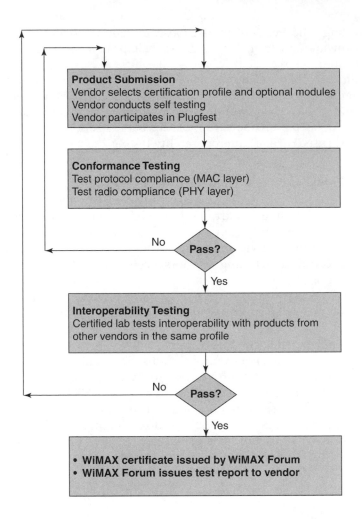

FIGURE 9.1
Summary of WiMAX certification process.

TABLE 9.1

The Role of Major Players in the WiMAX Certification Program

Body	Role	Institutions/Forums
Standards body	To develop standards specifications, test specifications, PICS, TSS/TP, ATS	IEEE, ETSI, WiMAX Forum
Regulator body	To establish policies and procedures for certification	WiMAX Forum
Certification bodies	To certify products	WiMAX Forum
Testing laboratories	To test products according to test specifications	CETECOM lab, Spain TTA, South Korea

TABLE 9.2

WiMAX Forum Certification Profiles Based on Fixed WiMAX (IEEE 802.16-2004, OFDM)

No.	Spectrum Band (GH$_2$)	Duplexing Mode	Channel Width (MH$_2$)
1	3.5	TDD	3.5
2	3.5	FDD	3.5
3	3.5	TDD	7
4	3.5	FDD	7
5	5.8	TDD	10

Source: From WiMAX Forum, http://www.wimaxforum.org.

WiMAX Forum defines a list of test cases to use during the certification process for all equipment based on the same system profile.

9.3.2.1 Certification Profiles

Certification profiles are instantiation of a system profile; that is, for each system profile there are multiple certification profiles. A certification profile is defined by three parameters:

- Spectrum band (<11 GHz)
- Channel width (size) (1.75–10 MHz)
- Duplexing type (TDD or FDD)

All certified products fully interoperate with other products in the same certification profile that is tested under the same release, and is backward compatible with products tested under previous releases. The definition of certification profiles depends on the market and vendor demand, which in turn is linked to spectrum availability in different countries. Vendor interest is clearly a prerequisite for interoperability, as a minimum of three vendors are needed to initiate interoperability testing.

For 802.16-2004 system profile, five certification profiles have been defined as shown in Table 9.2 [2].

It should be noted that the certification profiles define classes of interoperable equipment for testing purposes. The actual coverage of interoperability among various products will vary due to the fact that multimode BSs or SSs, may work at different frequencies or channel widths, or duplexing modes. For example, a multimode SS will interoperate with any BS that supports any of the certification profiles for which the SS is certified.

9.3.2.2 The WiMAX Forum Certification Process

WiMAX products need to pass two stages in the testing process to gain certification—compliance (or conformance) testing and interoperability testing. Compliance testing ensures that the product complies (or conforms) with the test specification set forth in the given system profile. Interoperability testing ensures that the subscriber units and the BSs from different vendors operate within the same network.

As the technology and user requirements change over time, the certification programs need to adapt to these changes. Therefore, the scope of certification expands over time with the addition of new test cases. The list of requirements as defined in the system profiles does not change to ensure backward compatibility. The new test cases are introduced either to include new features in the certification process or to expand the coverage of the existing ones.

WiMAX certification testing will be performed in multiple releases and waves. Multiple releases are necessary to test the products as they evolve over time with new features or specifications. For each release, the procedure will test all the new features as well as previously tested features. Each release will be backward compatible with the previous releases. For example, release 3 testing will include testing of all requirements from release 1 and release 2. For each release, there will also be different waves of testing, which will involve the same tests but at different times with different vendors. For example, vendors A, B, and C perform release 1 wave 1 testing in month 1, while vendors D, E, and F perform release 1 wave 2 testing in month 3 [6].

Release 1 for fixed WiMAX covers only the mandatory features and includes testing for the air interface, network entry, dynamic services, and bandwidth allocation. Release 2 will introduce the following three optional modules:

- Quality of service (QoS) support
- Advance security with advanced encryption standard (AES)
- Automatic repeat request (ARQ)

It is quite possible that a release 1 BS supports QoS, but this feature is not tested in release 1; therefore, interoperability, while possible, should not be expected.

9.3.3   Abstract Test Suite

The WIMAX Forum has developed several process and procedural test documents based on IEEE 802.16 standards. The key test documents are as follows:

- Protocol implementation conformance specifications (PICS)
- Test purposes and test suite structure (TP and TSS)
- Radio conformance test (RCT) specifications
- Protocol implementation extra information for testing (IXIT)

To evaluate conformance of a particular implementation, it is necessary to have a statement listing the capabilities and options that have been implemented for a telecommunication specifications. Such a statement is known as PICS.

The end product of the abstract test suite (ATS) are the test scripts for conformance and interoperability testing under a number of test conditions mentioned in the PICS document for a given WiMAX system profile.

9.3.3.1 Certification Challenges

One of the major challenges for the WiMAX certification process is to have all the necessary test equipment, such as a protocol analyzer (PA), ready for use by the certification lab. The WiMAX Forum is working on the development of a PA (through a third party) to help analyze the transmitted downlink and uplink IP packets between a BS and SS based on the PICS document [8]. Some key features of the PA system are:

- Data packet capture and display
  - Display multiple levels of information (summary, decode tree, raw data packets, etc.)
  - Time stamping
  - Ability to correlate capture data with test results
- Display and store test log
- Display of message sequence information
- Ability to trigger on packet contents (protocol, field values, pattern) and on extended sequences of events
- Collection and display of data statistics (packet type counters, error counter, traffic statistics, bandwidth utilization)
- Support of a scripting interface to create customized scripts and to automate testing process

9.4 WiMAX Plugfest Testing

The main idea of group tests or Plugfest is to provide vendors the opportunity to address potential problems at an early stage. The Plugfest is a preview of full interoperability testing before formal certification testing. It allows vendors to get an early look at how well their equipment interoperates. The Plugfest is a weeklong event carried out at a WiMAX Forum-contracted testing site. The participating vendors must first agree on a set of RF/PHY characteristics within a given certification profile, and a minimum of three vendors must be available to conduct interoperability testing. The WiMAX Forum certification working group (CWG) has indicated that the Plugfest events will be planned about every six months. In some cases, there may be additional Plugfest events planned to prepare vendors for future certification releases. We now describe the test architectures for the following Plugfests:

- 3rd Plugfest for fixed WiMAX (held in March 2006) [7]
- 5th Plugfest for mobile WiMAX (held in September 2006) [12]

9.4.1   3rd Plugfest Test Architecture

There are four system test configurations defined in the 3rd Plugfest. A system under test (SUT) is defined as a network consisting of one BS and 1–3 SSs and, when needed, some monitoring devices (such as PA, VSA). The following specific configurations are used [7]:

- SUT1: 1 BS + monitoring devices
- SUT2: 1 BS + 1 SS + monitoring devices
- SUT3: 1 BS + 2 SSs (from different vendors and at least one of them from a different vendor to that of the BS) + monitoring devices
- SUT4: 1 BS + 3 SSs (from different vendors and at least two of them from different vendors to that of the BS) + monitoring devices

The devices in each SUT will be interconnected by wired means (using, for example, attenuators, couplers, etc.). The certification profiles tested in 3rd Plugfest consisted of:

- 3.5 GHz, 3.5 MHz, TDD
- 3.5 GHz, 3.5 MHz, FDD

9.4.1.1   SUT Configuration 1

Figure 9.2 shows the SUT test configuration. The BS transmits broadcast messages while the monitoring device captures all the necessary MAC and PHY/RF parameters. The BS is connected to a test generating equipment/packet generator/controller via a local area network.

9.4.1.2   SUT Configuration 2

The test configuration 2 is shown in Figure 9.3. The BS transmits and receives data packets from a single SS while a monitoring device captures all the necessary MAC messages and PHY/RF parameters.

9.4.1.3   SUT Configuration 3

In this test configuration, shown in Figure 9.4, the BS transmits and receives data packets from two SSs (SS1 and SS2) while a monitoring device captures all the necessary MAC messages and PHY/RF parameters. SS1 and BS vendors are different, while SS2 may be from the same vendor as the BS.

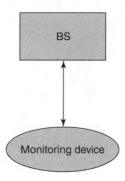

FIGURE 9.2
Setup for SUT1.

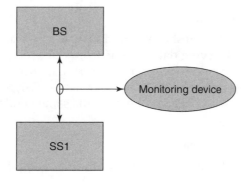

FIGURE 9.3
Setup for SUT2.

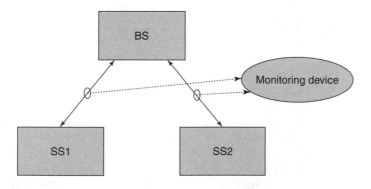

FIGURE 9.4
Setup for SUT3.

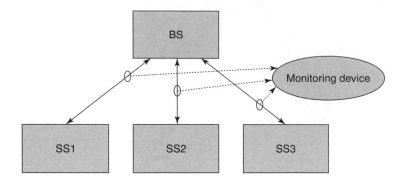

FIGURE 9.5
Setup for SUT4.

9.4.1.4   SUT Configuration 4

Figure 9.5 shows the SUT test configuration 4. In this configuration, the BS is transmitting and receiving data packets from three SSs (SS1, SS2, and SS3) from three different vendors, while a monitoring device is capturing all the necessary MAC messages and PHY/RF parameters. SS1, SS3, and BS belong to different vendors while SS2 may be from the same vendor as the BS.

9.4.1.5   Test Plan for Plugfest

Test cases for the Plugfest are organized into the following six different groups:

1. Radio link control
2. Initialization
3. Privacy and key management
4. Dynamic services
5. Bandwidth allocation and polling
6. Classification

Test cases are developed to test functionality in each of the above-mentioned group. For the radio link control, test cases are developed to test initialization, initial ranging, ability to negotiate basic capabilities, etc. For the initialization group, the test cases address the registration and IP connectivity functionalities. For the privacy and key management, test cases are developed to test authentication/authorization, encryption key transfer, and encryption and key scheduling. For the dynamic services group, the test cases test whether a dynamic service can be added, changed, and deleted. In the bandwidth allocation and polling group, the test case tests whether a allocation request can be made and granted correctly. In the classification group, the classification parameters with respect to Ethernet and IP layers are tested.

The test procedure for a given test case includes the following:

- *Test case ID*
- *WiMAX certification profile employed*
- *Test purpose*: Which specific functionality is being tested?
- *SUT configuration used*: SUT1, SUT2, ...
- *Pretest conditions*: What will be the state of BS and SSs before the given test is executed? e.g., BS is switched on; SS1 is authorized, ...
- *Test setup*: A logical diagram consisting of all networking and monitoring devices
- *Testing steps along with test description and verdict (pass/fail)*
- *Observations during test*: expected behavior; unexpected behavior
- *BS id, SS ids*

To ensure that all the important functionalities needed to achieve interoperability are considered, a total of 79 test cases were developed for the 3rd Plugfest.

9.4.2   5th Plugfest Architecture

Five system test configurations are defined in the 5th mobile Plugfest. An SUT is defined as a network consisting of one BS and 1–3 mobile stations (MSs). The system also includes monitoring devices (such as WiMAX PA or VSA). The following configurations are used in the Plugfest:

- SUT1: Single BS and single MS (from same vendor)
- SUT2: Single BS and single MS (from different vendors)
- SUT3: Single BS and two MSs (from same vendor)
- SUT4: Single BS and two MSs (from different vendors)
- SUT5: Single BS and three MSs (three different vendors)

Devices in each SUT are interconnected by wired means.

9.4.2.1   *SUT1: Single BS and Single MS (from Same Vendor)*

In this configuration, a single BS is connected to a single MS. This is an initial test configuration for all vendors prior to engaging in interoperability testing to verify the operation of their own equipment. Vendors without having a BS and an MS of their own will not be able to run this test and will move to other testing.

9.4.2.2   SUT2: Single BS and Single MS (from Different Vendor)

In this test configuration, the BS is transmitting and receiving data packets from a single MS. A monitoring device can be used to capture all the necessary MAC messages and PHY/RF parameters.

9.4.2.3   SUT3: Single BS and Two MS (from Same Vendor)

In this test configuration, the BS is transmitting and receiving data packets from two MSs. The two MSs are from the same vendor but different vendor than the BS1.

9.4.2.4   SUT4: Single BS and Two MS (from Different Vendors)

In this test configuration, a single BS is connected to two MSs and those MSs are from different vendors. Ideally, all three units should be from different vendors. However, one of the MS vendors may also be the BS vendor, depending upon the equipment availability.

9.4.2.5   SUT5: Single BS and Three MS (Three Different Vendors)

In this configuration, a single BS is connected to three MSs. Each of the MSs may be from different vendors, or two MSs may be from the same vendor, or one of the MS could be from the same BS vendor, depending upon the equipment availability.

The certification profiles tested in the 5th Plugfest are:

- 2.3–2.4 GHz, 5/8.75/10 MHz, TDD
- 2.496–2.69 GHz, 5/10 MHz, TDD
- 3.4–3.6 GHz, 5/7 MHz, TDD
- 4.935–4.990 GHz, 5 MHz, TDD

TABLE 9.3

Test Scenario Structure for Mobile Plugfest

Test Scenario	Major Functionality Tested
Network entry procedure	• MS(s) synchronize to BS • Ranging • Capabilities negotiation • Authentication (not used) • Registration
Traffic connections establishment	• Service flow provisioning • Service flow activation
User data transfer	• Downlink PING • Uplink PING

Source: From WiMAX Forum, http://www.wimaxforum.org.

Testing scenarios are organized into the following three groups:

- Network entry procedures
- Traffic connections establishment
- User data transfer

Table 9.3 shows the test scenario structure used at this Plugfest.

9.5 Radio Conformance Testing and Measurements

The overall WiMAX system performance largely depends on the RF characteristics of the WiMAX devices. These RF specifications are defined in both the IEEE 802.16-2004, IEEE 802.16 radio conformance test [10], and "WiMAX Certification" documents. This section describes the major testing and measurement procedures for the RF transmitter and the receiver.

The WiMAX transmitter requirements are defined in Sections 8.3.10 and 8.5.2 of IEEE 802.16-2004, and they include:

- 8.3.10.1 Transmit power level control
- 8.3.10.1.1 Transmitter spectral flatness
- 8.3.10.1.2 Transmitter constellation error
- 8.5.2 Transmit spectral mask (for unlicensed band operations)

Some other key transmitter measurements such as adjacent channel power ratio (ACPR), maximum output power, spurious, and harmonics are not defined in the standards but are left to the local regulations [11].

9.5.1 Transmitter Power Level Control

The transmitter power level control requirement means that the BS and SS must be able to adjust their output power over a defined range [11]. As WiMAX systems can be used for nonline-of-sight applications, gain control of the transmitter is necessary to adjust the output transmit level depending on the channel quality. According to 802.16-2004 standards, a BS is required to have a minimum adjustment range of 10 dB, where SS must have a minimum adjustment range of 30 dB for devices not supporting subchannelization and 50 dB for devices that support subchannelizations. Within these ranges, step sizes must be a minimum of 1 dB and relative accuracy of all steps less than 30 dB must be ±1.5 dB. Larger steps have a ±3 dB relative accuracy.

The recommended RF test equipment for test and measure transmitter power level control requirement may include a spectrum analyzer, average power meter, and VSA. For the measurement for transmit power level control, the device under test (DUT) is set to various output power settings. The

DUT is set to transmit at valid output power with a frame structure that has a proper preamble and data burst. The preamble symbols are transmitted at a 3 dB higher power than the data bursts. The DUT output power of data burst is measured with the test equipment.

9.5.2   Transmitter Spectral Flatness

During normal system operation, all uplink and downlink transmissions in WiMAX begin with a preamble. The preamble is used to synchronize the receiver with the transmitter and perform various channel estimation and equalization processes. The preamble uses QPSK modulation and has no embedded BPSK pilots. Owing to this property, the preamble is ideally suited to specify the spectral flatness across all the subcarriers. According to the 802.16-2004 standard, the data shall be taken from the "channel estimation step" (which happens to be the preamble), and the absolute difference between adjacent subcarriers shall not exceed 0.1 dB.

For the measurement of transmitter spectral flatness, the DUT is set to transmit at valid output with a frame structure that has a proper preamble. The DUT spectral flatness is measured with the test equipment (VSA or spectrum analyzer) and the relative amplitude of each subcarrier with adjacent subcarriers is compared [11].

9.5.3   Transmitter Constellation Error

Transmitter constellation error test is a measurement of the transmitter modulation accuracy. Accurate transmitter modulation is necessary to ensure that the receiver can demodulate the signal with minimal decode errors [11]. This measurement is quite similar to error vector magnitude (EVM) used in the industry for other digital communications technologies. The IEEE 802.16 standard introduced a new term called relative constellation error (RCE), which determines the magnitude error of each constellation point and RMS (root mean squared) averages them across multiple symbols, frames, and packets. Each of the seven burst profiles (modulation/coding types) used in WiMAX has a specification for RCE expressed in dB. For example, RCE for the profile (QPSK, 1/2) is −16.0 dB.

For the measurement of RCE, the DUT is set to transmit valid subframes (preambles followed by data bursts). The DUT RCE is then measured with test equipment (such as spectrum analyzer with support for WiMAX analysis). The above procedure is repeated for various band edges for RF frequencies, modulation types, and power levels.

9.5.4   Transmitter Spectral Mask

This test measures the spectral profile of the transmitter to verify that the device is not transmitting excessive energy outside its assigned channel bandwidth. As local regulations will control the WiMAX frequency bands,

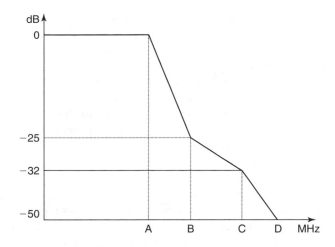

FIGURE 9.6
Transmitter spectral mask.

TABLE 9.4

Spectral Mask Limits Used for Figure 9.6

Channelization (MHz)	A	B	C	D
20	9.5	10.9	19.5	29.5
10	4.75	5.45	9.75	14.75

the IEEE 802.16 specification does not attempt to specify the spectral mask for any operation other than the unlicensed band [11]. Figure 9.6 and Table 9.4 show the spectral mask specification for 802.16 unlicensed band operations.

For the measurement of spectral profile, the DUT is set to transmit valid subframes (preambles followed by data bursts). The DUT spectral output is measured using the test equipment (e.g., spectrum analyzer). The above procedure is repeated for various RF frequencies, modulation types, and power levels.

9.5.5 Nonspecified Transmitter Measurements

The ACPR (sometimes also known as adjacent channel leakage ratio, or ACLR) is a measure of the transmitter energy that is leaking into an adjacent or alternate channel [11]. In practice, a very small amount of the transmitter energy will show up in other nearby channels. A spectrum analyzer can easily make this measurement. First, we can measure the in-channel power within the assigned channel for the DUT. The spectrum analyzer can be retuned to a frequency offset one channel away and the leakage power is measured. ACPR is the ratio of average power in the adjacent frequency channel to the average power in the assigned channel. The acceptable value for ACPR is from 30 to 80 dB, depending upon the application [11].

The maximum output power for WiMAX applications is generally specified by local regulations, depending upon the band of operation. Higher transmitter output power will cause unnecessary interference in the system. Moreover, for handheld devices excessive output power will cause unnecessary battery drain. Test equipment such as power meters and spectrum analyzers are ideal for measuring output power.

A typical RF output stage of the transmitter will have some filtering mechanism to suppress unwanted signals from being transmitted. These unwanted signals can be classified as either harmonics or spurious signals. Harmonics are integer multiples of the primary transmitter frequency and therefore the frequency at which they appear is very predictable. Spurious signals are typically image frequencies caused by internal mixing of an oscillator or clock frequency with the primary transmitter output frequency. Spectrum analyzers are ideal instruments to be used for both harmonics and spurious measurements. Typically, harmonics are measured to at least the 5th harmonic [11].

9.5.6 Receiver Tests

The WiMAX receiver requirements are defined in Section 8.3.11 of IEEE 802.16-2004. The receiver tests include receiver sensitivity, receiver adjacent and alternate channel rejection, receiver maximum input signal, receiver maximum tolerable signal, and receiver image rejection.

The test for receiver sensitivity measures the receiver's performance using known signal conditions that include the modulation and coding rate, SNR, and input level. Using the specified conditions (given in IEEE 802.16-2004, Section 8.3.11), the receiver must be able to decode data bits with a bit error rate (BER) less than 10^{-6} after forward error correction (FEC). The standard specifies the test conditions for a variety of bandwidths and modulation types.

To measure receiver sensitivity, the RF signal generator is set (using software methods) to generate the test conditions defined in the standards. Any connector and cable losses between the RF source output and the DUT input must be compensated by adjusting the RF output level at the generator. The DUT is set to receive and decode a continuous stream of packets that contain the special data patterns defined in the standards. The DUT must somehow calculate BER or provide the data bits externally to the BER test set that can calculate BER by comparing the received data bits with the expected values. The BER calculation is done on fully decoded payload data that does not contain FEC. The above procedure is repeated for all valid modulation and coding types.

The other receiver tests can be performed as variations of the receiver sensitivity test described above.

9.6 Conclusions and Summary

The WiMAX technology offers unique testing challenges owing to its complex nature. It is expected that a huge proportion of budget will be spent on the testing activities of WiMAX products. The WiMAX products need to conform to the standards (IEEE 802.16, HiperMAN) and they need to interoperate with equipment from other vendors. Both conformance and interoperability of WiMAX products are crucial. The IEEE standards outline the conformance testing (including radio conformance testing) procedures. The WiMAX Forum is taking testing to another extreme by offering certification testing, which combines conformance and interoperability testing of the products. This chapter surveys the methodologies used in conformance, interoperability, and radio conformance testing. It also describes the WiMAX certification process and testing scenarios at recently held WiMAX Forum Plugfest events.

References

1. European Telecommunications Standards Institute (ETSI), http://www.etsi.org.
2. WiMAX Forum, http://www.wimaxforum.org.
3. A. Moreno, *WiMAX Certification*, http://www.cetecom.es.
4. Senza Fili Consulting, *The WiMAX Forum Certified Program for Fixed WiMAX*, http://www.wimaxforum.org, May 2006.
5. M. Paolini, *The Evolution of WiMAX Certification*, Senza Fili Consulting, http://www.senzafiliconsulting.com, October 2005.
6. WiMAX Forum Certification of Broadband Wireless Systems, Report prepared by Redline Communications on behalf of WiMAX Forum, http://www.wimaxforum.com, 2005.
7. 3rd Plugfest—Sophia Antipolis, France, March 2006, White Paper by Redline Communications, http://www.redlinecommunications.com.
8. E. Agis, Global, interoperable broadband wireless networks: Extending WiMAX technology to mobility, *Intel Technology Journal*, Vol. 8, Issue 3, 2004.
9. *IEEE Standard for Local and Metropolitan Area Networks—Part 16: Air Interface for Fixed Broadband Wireless Access Systems* (IEEE standard 802.16-2004).
10. *IEEE Standard 802.16 Conformance—Part 3: Radio Conformance Tests (RCT) for 10–66 GHz WirelessMAN-SC Air Interface*, 2004.
11. WiMAX Concepts and RF Measurements: IEEE 802.16-2004 WiMAX PHY Layer Operation and Measurements, Application note from Agilent Technologies, http://www.agilent.com.
12. 5th Plugfest, Bechtel Labs, MD, USA, September 2006, White paper available at WiMAX Forum website, http://www.wimaxforum.org.
13. L. Frenzel, Virtual RF design and testing, *Electronic Design*, 11 May, 2006.

Part II

Security

10

An Overview of WiMAX Security

Eduardo B. Fernandez and Michael VanHilst

CONTENTS

10.1 Introduction

The IEEE 802.16 protocol is also called WiMAX, which stands for worldwide interoperability of microwave access. It addresses high-bandwidth wide-area access between a service provider base station (BS) and multiple subscriber stations (SSs), often referred to as the "last mile" in reference to neighborhood connections between subscribers' homes and a phone or cable company office. In fact, important parts of the protocol are based on the DOCSIS BPI + (data over cable service interface specifications: baseline privacy plus interface specification) [3] protocol used in cable modems. The original 802.16 standard covers line-of-sight connections in the 10–66 GHz range, supporting speeds up to 280 Mbps over distances up to 50 km (30 mi.). 802.16a covers nonline-of-sight connections in the 2–11 GHz range, supporting speeds up to 75 Mbps over distances of 5–8 km (3–5 mi.). 802.16a also adds features for mesh networks, while the 802.16e standard adds support for mobility (i.e., station handoff) [4].

As a wireless protocol, WiMAX has an additional set of security threats not faced in cable systems. Because the DOCSIS protocol was developed for cable modems, not wireless systems, the original 802.16 standard does not provide enough security for the intended purpose. The standard threats for wireless systems still apply to WiMAX systems, in particular all the attacks to the higher levels [5,6]. Later extensions correct some of the weaknesses.

We present here an overview of the security aspects of this standard. We use unified modeling language (UML) class and sequence diagrams to describe architectural aspects. These are conceptual diagrams, intended to define the information in each unit and do not reflect implementation details. The reader is referred to Ref. 9 or similar textbooks for introductions to UML. The idea is to present a high-level overview that can be read before getting into the details of the standard or more advanced discussions.

Section 10.2 discusses the networking aspects of this protocol and Section 10.3 presents an overview of its basic security aspects. We end with some conclusions and ideas for future work.

10.2   Network Aspects

WiMAX/802.16 defines two layers of the protocol stack, physical (PHY) and medium access control (MAC). The MAC layer manages connections and security. The PHY handles signal connectivity and error correction, as well as initial ranging, registration, bandwidth requests, and connection channels for management and data. The PHY layer consists of a sequence of equal-length frames transmitted through the coding and modulation of RF signals. Physical frames, and also MAC frames, do not necessarily begin or end on boundaries of higher layer frames—this is handled by intermediate mapping layers. Intermediate mapping gives 802.16 flexibility to support a wide variety of traffic types and profiles in the transport layer and above, including IP, Ethernet, and ATM, with a high level of efficiency [4].

*SS*s communicate with a *BS* through wireless links. Before connecting, a SS scans its frequency list to find a BS, observes BS traffic to determine parameters for timing, modulation, error correction, and power, and finally identifies time slots (maintenance windows) to use for an initial request. The initial sequence of packets (ranging requests) between the SS and BS are used to refine power and timing settings, and to establish connection reservations (time slot profiles and connection IDs or CIDs). The SS obtains multiple CIDs for different management and data connections with different quality of service (QoS) criteria. Subsequent management messages can change connection profiles in response to changing QoS needs and signal quality.

A communication is divided into frames. Frames from BS to SS (downlink frames) and SS to BS (uplink frames) contain a frame header and a body (Figure 10.1). The header has two slot maps, a downlink map (DL_MAP) and an uplink map (UL_MAP). The maps describe the use of the slots and their location. Each slot is part of some connection, identified by a CID. Management connections are used to set up connections and contain aspects such as bandwidth requests and other administration information. On connection, an SS is assigned three management connections (basic, primary, and secondary) for management messages with different QoS needs. Short

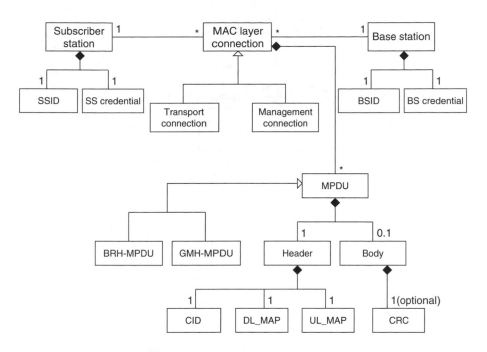

FIGURE 10.1
Class diagram of WiMAX network architecture.

management messages needing immediate response use the basic connection, while the secondary connection handles IP management traffic such as address request (DHCP), system status (SNMP), and remote update (TFTP). User messages are sent through transport connections. IEEE security applies only to transport connections and the secondary management channel.

Data is moved through packets with MAC protocol data units (MPDUs). Depending on their functions there are two types of MPDUs (Figure 10.1): those with bandwidth request headers (BRHs) and those with generic MAC headers (GMHs) (in this case the header is followed by a body and an optional Cyclic redundancy code (CRC)). A management connection uses management packets, where each MPDU carries a single MAC management message.

10.3 WiMAX Security

802.16 defines a privacy and key management (PKM) protocol to address the goals of SS privacy and preventing theft of provider services [2]. What they really mean is confidentiality and key management. Privacy is the right of individuals to control information about themselves [7], while confidentiality (secrecy) is the restriction where users cannot read information

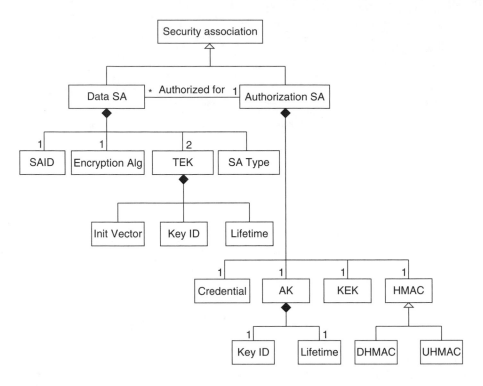

FIGURE 10.2
Class diagram of SA structure.

without authorization, which is clearly the case here. The PKM uses security associations (SAs) of which there are two types. A data SA specifies how messages between the BS and SS are to be encrypted, which algorithms will be used, the keys to be used, and related information. By using additional SAs, different methods of encryption may be used for different groups of messages. Each data SA includes an ID (SAID), an encryption algorithm to protect the confidentiality of messages, two traffic-encryption keys (TEKs), two identifiers (one for each TEK), a TEK lifetime, an initialization vector for each TEK, and an indication of the type of data SA (primary or dynamic). An authorization SA (not explicitly defined by the standard) includes a credential, an authorization key (AK) to authorize the use of the links, an identifier for the AK, a lifetime for the AK, a key-encryption key (KEK), a downlink hash-based message authentication code (DHMAC), an uplink hash code (UHMAC), and a list of authorized data SAs. Figure 10.2 summarizes the information used in SAs.

Security is closely tied to connections and connection types. WiMAX defines two connection types, management and data. As indicated earlier, management connections are further subdivided into basic, primary, and secondary.

Security begins with authentication in the initial ranging request phase. Each SS has a 48-bit ID (or MAC address) and an X.509 certificate. It also

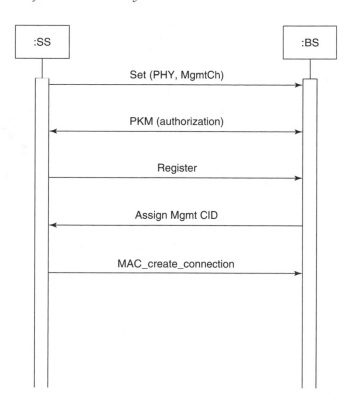

FIGURE 10.3
Starting a connection.

possesses an X.509 certificate of its manufacturer—but this latter certificate is generally ignored by the BS and plays no role in security. Figure 10.3 is a sequence diagram of how an SS starts to use (enters) the network. After the SS finds a BS downlink signal, the SS sets up its PHY parameters and establishes a management channel that can be used for further negotiation. It then starts an authentication protocol (PKM authorization, described later in Figure 10.5). The SS registers itself with the BS by sending a registration request. The BS responds with a registration reply in which the SS is assigned a channel ID for a secondary management channel. After that, the SS creates a transport connection through the BS using a MAC_create_connection request.

Stations perform authentication using credentials, X.509 certificates in the current standard. Figure 10.4 shows a class diagram to describe the structure of these certificates. Once authenticated, a user is given a token to access the system. Figure 10.5 summarizes the steps in the PKM protocol for the SS to obtain authorized access to the network. The SS sends two messages. The first message contains the manufacturer X.509 certificate. The second, authorization request, includes its own X.509 certificate and a list of its security capabilities. If the SS is authenticated and authorized to join the network, the BS sends an authorization reply. The authorization reply is encrypted with

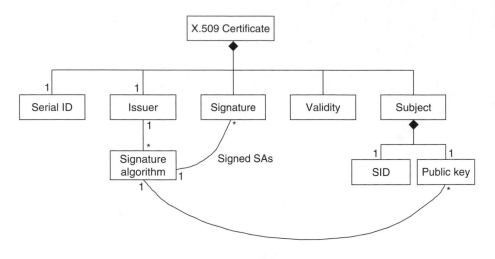

FIGURE 10.4
Class diagram for X.509 certificates.

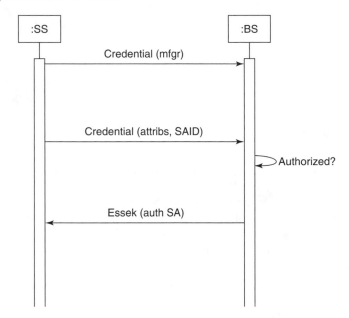

FIGURE 10.5
PKM authorization protocol.

the SS's public key (denoted as Essek in the figure) and includes an AK, a key lifetime, a key sequence number, and an SA descriptor (the basis for the authorization SA).

The PKM exchange of messages establishes an authorization key and a SA. The sequence numbers in the protocol represent instances of the AK. The AK is used to derive three additional keys for both encrypting and

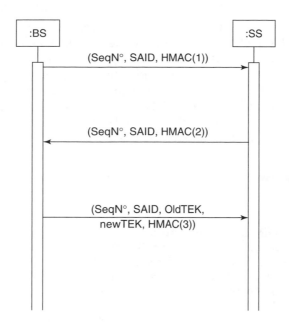

FIGURE 10.6
Creation of a data association.

verifying the source and integrity of future messages. Message source and integrity are verified with message authentication (HMAC) keys, e.g., HMAC(1) proves the integrity of the first message from the BS to the SS. Two separate HMAC keys are derived from the AK, for the BS-to-SS (downlink) and SS-to-BS (uplink) directions. A KEK is also derived from the AK. The KEK is used for key exchange messages to obtain the transmission encryption keys used when transmitting data.

In WiMAX, the SS and BS exchange management messages for authentication as shown and then proceed to key management (as shown in Figure 10.6) before transmitting data. The old TEK is the one currently used; it will be replaced by the new TEK at its expiration. Chapters 11 and 12 will discuss known issues in the different standards that govern keys and authentication.

Several flaws have been found in this standard. Refs. 1,8,11 discuss some of them. An improved scheme for key management, based on extensible authentication protocol (EAP), is presented in Ref. 12. Other improvements and issues are discussed in Chapters 11 and 12 in this book.

10.4 Conclusions

We have distilled the fundamental aspects of the conceptual architecture of WiMAX, in particular its security in the form of UML models. We have separated the conceptual architecture from implementation details, aspects that are intermingled in the standards. This separation is very important for evolving standards like this one, where the implementation is expected to change relatively frequently, but the conceptual architecture should remain stable. These models can be used to understand the more complex aspects of the standard and to analyze weaknesses and improvements to the protocol. We are now converting these models into security patterns in the style of Ref. 10. In the form of patterns, these models can be used to guide the design of wireless systems and to compare standards.

References

1. M. Barbeau, WiMAX/802.16 threat analysis, in *Proceedings of ACM Q2SWinet'05*, Montreal, Quebec, Canada, October 13, 2005.
2. J.L. Burbank and W.T. Kasch, IEEE 802.16 broadband wireless technology and its application to the military problem space, in *Proceedings of the 2005 IEEE Military Communications Conference (MILCOM 2005)*, Vol. 3, pp. 1905–1911, October 2005.
3. *Data-Over-Cable Service Interface Specifications: Baseline Privacy Plus Interface Specification SP-BPI+-I05-000714*, DOCSIS, http://www.cablemodem. com/, July 2000.
4. C. Ecklund, R.B. Marks, K.L. Stanwood, and S. Wang, IEEE Standard 802.16: A technical overview of the wirelessMAN air interface for broadband wireless access, *IEEE Communications Magazine*, Vol. 40, No. 6, pp. 98–107, June 2002.
5. E.B. Fernandez, S. Rajput, M. VanHilst, and M.M. Larrondo-Petrie, Some security issues of wireless systems, in *Proceedings of IEEE Fifth International Symposium and School on Advanced Distributed Systems (ISSADS 2005)*, Guadalajara, Mexico, January 24–28, 2005, and in F.F. Ramos et al. (eds.) *Lecture Notes in Computer Science*, LNCS Vol. 3563, Springer-Verlag, Berlin, Heidelberg, pp. 388–396, 2005.
6. E.B. Fernandez, I. Jawhar, M.M. Larrondo-Petrie, and M. VanHilst, An overview of the security of wireless networks, Chapter 3 in the *Handbook of Wireless LANs*, M. Ilyas and S. Ahson (eds.), CRC Press, Boca Raton, FL, pp. 51–68, 2005.
7. D. Gollmann, *Computer Security* (2nd ed.), West Sussex, Wiley, 2006.
8. D. Johnston and J. Walker, Overview of IEEE 802.16 security, *IEEE Security and Privacy*, pp. 40–48, May/June 2004.
9. C. Larman, *Applying UML and Patterns: An Introduction to Object-Oriented Analysis and Design and Iterative Development*, 3rd edition, Prentice-Hall, Upper Saddle River, NJ, 2005.
10. M. Schumacher, E.B. Fernandez, D. Hybertson, F. Buschmann, and P. Sommerlad, *Security Patterns: Integrating Security and Systems Engineering*, Wiley, Chichester, W. Sussex, England, 2006.
11. S. Xu, M. Mathews, and C.T. Huang, Security issues in privacy and key management protocols of IEEE 802.16, in *Proceedings of ACM SE'06*, Melbourne, FL, March 2006.
12. F. Yang, H. Zhou, L. Zhang, and J. Feng, An improved security scheme in WMAN based on IEEE Standard 802.16, *Wireless Communication, Networking and Mobile Computing*, Vol. 2, pp. 1191–1194, September 2005.

11

Privacy and Security in WiMAX Networks

Amitabh Mishra and Nolan Glore

CONTENTS

11.1   Introduction

IEEE has created a new standard called IEEE 802.16 that deals with providing broadband wireless access to residential and business customers and is popularly known as the worldwide interoperability for microwave access (WiMAX). WiMAX is a nonprofit industry trade organization that oversees the implementation of this standard. WiMAX is expected to replace expensive services like cable, DSL, and T1 for last mile broadband access because it has a target transmission rate that can exceed 100 Mbps. The transmission range for WiMAX devices is up to 31 mi., which also far exceeds WiFi's transmission range of approximately 100 m [1,2]. With such a large transmission range, a single base station may be able to provide broadband connections to an entire city.

WiMAX was designed with the ability to provide quality of service (QoS) in mind's as a result, it can support delay-sensitive applications and services. Since it is connection-oriented, it has the ability to perform per connection QoS, allowing it to operate in both dedicated and best effort situations.

WiMAX was created to meet the growing demand for broadband wireless access (BWA). This demand has proven to be challenging for service providers due to the absence of a global standard. Currently, many service providers have created proprietary solutions based on a modified version of 802.11. Unfortunately these are costly solutions, which do not offer compatibility or flexibility. Some providers have tried to use 802.11 to implement a citywide deployment, despite the fact that it was designed to connect home or office computers over short distances.

When current WLAN technologies were examined for outdoor applications, it became clear that WiFi was not well suited for outdoor BWA applications or to provide T1 level access to businesses. A technology was needed that could operate in an outdoor environment and provide T1 level services to support data, voice, video, wireless backhaul for hotspots, and cellular tower backhaul services. IEEE 802.16 standard was created in response to support these services, and while this standard was being defined a major emphasis was placed on the physical (PHY) layer to support an outdoor environment and on the media access control (MAC) layer to provide QoS for delay sensitive applications.

11.1.1   Standardization and Certification

A group of industry leaders (Intel, AT&T, Samsung, Motorola, Cisco, and others) have been chartered to promote the adoption of WiMAX. Together they make up the WiMAX Forum, which has developed a certification program for WiMAX-enabled devices [1]. The goal of the forum is to define and conduct interoperability testing and award "WiMAX Certified™" labels to vendor systems that pass these tests. The approach is similar to the one taken by the WiFi Alliance, which helped bring wireless LANs to the masses. The WiMAX certification process will also consider the European Telecommunications Standards Institute's MAN standard (HiperMAN), which will allow WiMAX-certified devices to work in both the United States and Europe. The HiperMAN and 802.16 are both being modified in a way such that they share the same PHY and MAC layers [1].

11.1.2 Frequencies

The initial 802.16 standard specifies operation frequencies between 10 and 66 GHz. The advantage of using these high frequencies is that they have more available bandwidth and less risk of interference. The disadvantage is that they require line-of-sight (LOS) environments. The 802.16a standard was adopted to provide operation in the 2 to 11 GHz frequency band. The use of these lower frequencies provides the ability to support nonline-of-sight (NLOS) operation [1].

Initial WiMAX deployments are expected to use the 5 GHz (license-exempt) and 2.5 GHz (licensed) frequency bands. Bands between 5.25 and 5.28 GHz will be the focus for rural areas with a low population density. For fixed wireless access, most countries have allocated the bands between 3.4 and 3.6 GHz but the United States, Mexico, Brazil, and some Southeast Asian nations, have chosen instead the bands between 2.5 and 2.7 GHz. There are also bands of interest smaller than 800 MHz, which are currently vacant or used for analog TV, due to their ability to penetrate obstacles and propagate further.

WiMAX will support flexible channel sizes, which will provide the ability to meet the many different channel size requirements and frequency bands from around the world. It also defines a dynamic frequency selection scheme, which helps minimize interference and increase performance.

11.1.3 Modes of Operation

WiMAX was designed to support both point-to-point (P2P) and point-to-multipoint (PMP) topologies. While P2P can be used to support wireless network backbones, PMP is what the standard was mainly designed for. In a PMP scenario a base station distributes traffic to many subscriber stations. To yield a high efficiency, WiMAX uses a scheduling method in which base stations can only transmit in their time-slots and do not contend with one another. This works quite efficiently because, unlike 802.11 hotspots, which usually have bursty traffic, stations can aggregate traffic from several computers, producing a steady flow. WiMAX also supports a mesh mode of operation, allowing service providers to use NLOS operation by having subscriber stations communicate directly to each other and relay traffic. Figure 11.1 illustrates the use of mesh mode in WiMAX to provide NLOS service to residential customers [1].

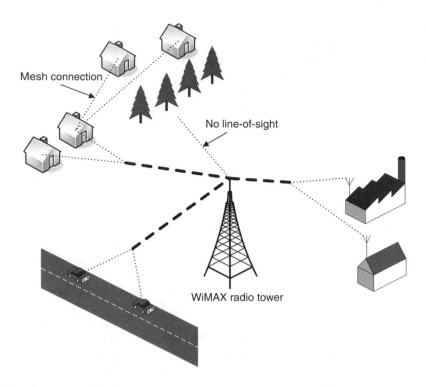

Mesh connection

No line-of-sight

WiMAX radio tower

FIGURE 11.1
WiMAX overview.

WiMAX's design allows it to be used in many different operating environments. The ability to provide last mile broadband access to consumers was one major consideration during development. With a focus on standardization and interoperability, WiMAX may provide a low-cost solution. Figure 11.1 illustrates the possible uses of WiMAX, including reliable business access, residential access, and high-speed connectivity for mobile users.

WiMAX's ability to provide high transfers rates allows it to be used as a network backbone. Specifically, developers envision using WiMAX as a backbone for 802.11 hotspots to provide Internet access. In this configuration, users would connect to a nearby 802.11 base station. The base station would then relay the user's data to a central WiMAX base station, which is connected to the Internet. This would provide citywide Internet access without having to run cables to each 802.11 hotspot.

Another access method would be to allow users to directly connect to the WiMAX base station, allowing citywide Internet access with a single point of attachment, without the need for any 802.11 base stations. Although possible, it may not use bandwidth as efficiently as the previous example. This is due to the scheduling algorithm WiMAX uses, which is designed for steady and smooth traffic and not for the bursty traffic created by individual users.

Also, it is likely that base stations may only have a range of 5–6 mi. instead of 30 mi. due to the increased vulnerability of the links from the user mobility.

Developers also had cellular applications in mind when they designed WiMAX. The first use of WiMAX for cellular applications will be a tower backhaul service. Once the IEEE 802.16e standard is implemented, which is optimized to support handoffs and roaming at speeds of up to 75 mph, WiMAX can be used to connect directly to cell phones and other mobile devices.

Many government agencies see the value of using WiMAX for both homeland security and its use in emergency situations. Agencies could deploy WiMAX-enabled devices to monitor high-value infrastructures and transmit the information to a central operations center for processing [3]. Emergency mobile wireless networks are another important use for the government. During a disaster, where all communications have been lost, a WiMAX network could be quickly set up. This would allow organizations like FEMA, Red Cross, and NATO to communicate important information that may be crucial to rescue operations [4].

11.2   Frame Structure

11.2.1   Physical Layer

When the 802.16 standard was introduced, it had a single carrier (SC) PHY specification to support LOS operations in the 10–66 GHz frequency band. With the 802.16a amendment to the standard, changes to the PHY were needed to support the 2–11 GHz frequency band. This led to the introduction of a new SC PHY, a 256 point FFT OFDM PHY, and a 2048 point FFT OFDMA PHY. The 802.16e amendment to the standard provides an enhanced version of OFDMA called scalable OFDMA (SOFDMA) [1].

The SC PHY specification is designed for LOS operation in the 10–66 GHz frequency band. Both TDD and FDD configurations are supported to allow for flexible spectrum usage. The SC PHY is designed for NLOS operation in the 2–11 GHz frequency band and is based on SC technology. The OFDM PHY uses a 256-carrier OFDM and TDMA to provide multiple access to different subscriber stations. The OFDMA PHY uses a 2048-carrier OFDM design. Multiple access is provided by assigning a subset of the carriers to an individual subscriber station. The enhancement of the OFDMA PHY, SOFDMA, uses the values 128, 512, 1024, and 2048 to scale the number of subcarriers in a channel [1,5].

The WiMAX Forum decided that the first interoperable tests and certifications for 802.16 devices would support OFDM. While OFDMA can allocate spectrum more efficiently and reduce interference, compared to OFDM it is more complex to install and operate. Therefore, OFDMA is only required for

802.16e certified devices, where it is needed to support mobile customers. The WiMAX Forum has worked with the Korean standard WiBro, which uses SOFDMA to insure the two technologies will be interoperable. Eventually, SOFDMA will be the PHY layer of choice for indoor and mobile equipment [1].

11.2.2   MAC Layer

In a WiMAX environment, it can be difficult for subscriber stations to listen to one another. Therefore, the MAC layer was designed to use a flexible frame structure in which the base station schedules subscriber station transmissions in advance. This reduces contention because subscriber stations only need to contend when they access the channel for the first time. The overall effect is increased efficiency, which allows one base station to serve a large number of subscriber stations.

11.2.2.1   MAC Protocol Data Unit

Each protocol data unit (PDU), as seen in Figure 11.2, is comprised of a generic MAC header (GMH), a payload, and an optional cyclic redundancy check (CRC). The GMH defines the contents of the payload and starts at the most significant bit (MSB). The payload consists of zero or more subheaders and MAC service data units (SDUs). The length of the payload may vary. The CRC is optional in the SC PHY layer but mandatory for SCa, OFDM, and OFDMA PHY layers [5].

There are two formats defined for the MAC header. The GMH is used for MAC PDUs that contain MAC management messages or convergence sublayer data. The bandwidth request header is used when requesting additional bandwidth. The two headers are distinguished by the single-bit header type (HT) field, which is zero for the generic header and one for the bandwidth request header.

The GMH, shown in Figure 11.3, is encoded from the HT field on. It is 6 bytes in length and consists of 12 fields. Two of these fields, which are 1 bit in length each, are reserved for future use. The remaining fields are defined in Table 11.1.

The type field of the GMH is used to indicate what type of subheader or special payload is included in the message. The possible type values and corresponding meanings are defined in Table 11.2.

FIGURE 11.2
MAC PDU.

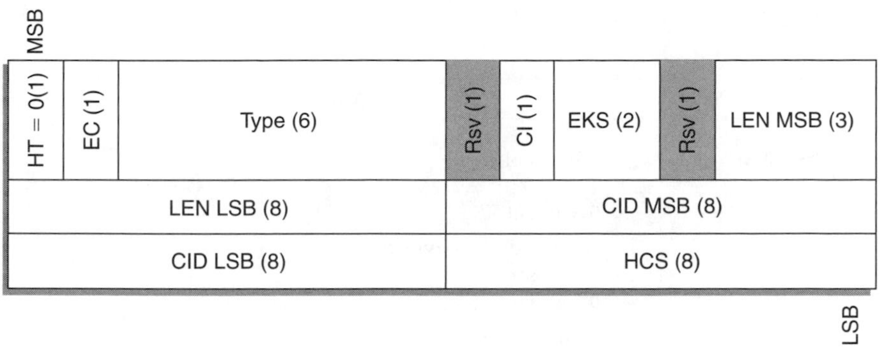

FIGURE 11.3
Generic MAC header.

TABLE 11.1

Generic MAC Header Fields

Name	Length (Bits)	Description
CI	1	CRC indicator 1 = CRC is included in the PDU by appending it to the payload after encryption if any 0 = No CRC is included
CID	16	Connection identifier
EC	1	Encryption control 0 = Payload is not encrypted 1 = Payload is encrypted
EKS	2	Encryption key sequence The index of the traffic encryption key (TEK) and initialization vector used to encrypt the payload. This field is only meaningful if the EC field is set to 1
HCS	8	Header check sequence An 8-bit field used to detect errors in the header
HT	1	Header type Shall be set to zero
LEN	11	Length The length in bytes of the MAC PDU including the MAC header and the CRC if present
Type	6	This field indicates the subheaders and special payload types present in the message payload

The bandwidth request PDU, shown in Figure 11.4, has no payload and consists of only the header. It is 6 bytes in length and consists of 8 fields, which are defined in Table 11.3. Like the GMH, the bandwidth request header is encoded from the HT field on.

TABLE 11.2

Type Encodings

Type Bit	Value
#5 (MSB)	Mesh subheader 1 = present 0 = absent
#4	ARQ feedback payload 1 = present 0 = absent
#3	Extended type Inidicates whether the present packing or fragmentation Subheaders are extended 1 = extended 0 = not extended
#2	Fragmentation subheader 1 = present 0 = absent
#1	Packing subheader 1 = present 0 = absent
#0 (LSB)	Downlink: FAST-FEEDBACK allocation subheader Uplink: grant management subheader 1 = present 0 = absent

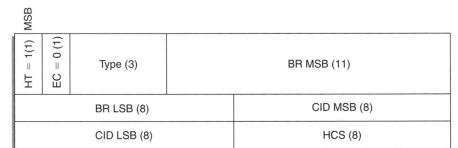

FIGURE 11.4
Bandwidth request header.

11.2.2.2 PMP

In PMP mode, multiple subscriber stations connect to a single base station. Each subscriber station is uniquely defined by a 48-bit universal MAC address. It is used during the initial ranging process and during the authentication process so the base station and subscriber station can verify each other's identity [5].

TABLE 11.3

Bandwidth Request Header Fields

Name	Length (Bits)	Description
BR	19	Bandwidth request The number of bytes of uplink bandwidth requested by the subscriber station. The bandwidth request is for the CID. The request shall not include any PHY overhead
CID	16	Connection identifier
EC	1	Always set to zero
HCS	8	Header check sequence An 8-bit field used to detect errors in the header
HT	1	Header type = 1
Type	3	Indicates the type of bandwidth request header

When a subscriber station first connects to a base station, two pairs of management connections are created between the subscriber station and the base station. An optional third pair of management connections may be created. Each pair consists of one uplink and one downlink connection, identified by a 16-bit connection ID (CID). Short, time-urgent management messages are sent over the basic connection. Longer, delay-tolerant management messages are sent over the primary management connection. Standards-based (i.e., DHCP, TFTP, SNMP) messages are sent using the secondary management connection.

Base stations do not have to coordinate their transmissions with other stations. They simply divide time into uplink and downlink transmission periods using TDD. Downlink messages are generally broadcast. A downlink map (DL-MAP) message can be used to define access to the downlink information by defining burst start times to subscriber stations. If a DL-MAP does not explicitly indicate a portion of the downlink for a specific subscriber station, then all subscriber stations capable of listening will listen. The subscriber stations will check the CIDs of the PDU and keep only the ones addressed to them.

Uplink transmissions to the base station are shared among subscriber stations and are on a demand basis. Subscriber stations use an uplink map (UL-MAP), which is obtained from the base station, to determine when it can transmit. Four different types of uplink scheduling mechanisms are used to control contention between users and tailor the delay and bandwidth requirements of each user application. These are implemented using unsolicited bandwidth grants, polling, and contention procedures. Performance can be optimized by using different combinations of these bandwidth allocation techniques.

11.2.2.3 Mesh

In mesh mode, subscriber stations can transmit to each other directly, allowing traffic to be routed through subscriber stations if two nodes cannot communicate directly. The advantage of mesh mode is it can provide NLOS communication for stations using higher frequency bands. This is accomplished by marking a node as a mesh base station if it has a direct connection to backhaul services outside the mesh network. Otherwise it is marked as a mesh subscriber station. Traffic can then flow from mesh subscriber stations to mesh base stations, then out of the mesh network and vice versa [5].

Similar to PMP mode, each node is uniquely defined by a 48-bit universal MAC address. It is used during the network entry process and the authentication process where the entry node and the network verify each other's identity. Once a node is authorized to the network, it requests a 16-bit node identifier (node ID) from the mesh base station. This node ID is used to identify nodes during operation.

Nodes view other stations in its mesh network in three different ways. Neighbors are stations to which the node has a direct link, which are considered to be "one hop" away. A neighborhood consists of all the neighbors of a node. Finally, an extended neighborhood contains all the neighbors of the neighborhood in addition to the neighborhood itself.

All communications within a mesh network are in the context of a link. 8-bit link identifiers (link IDs) are used to address nodes in the local neighborhood. Each link established between a node and its neighbor shall be assigned a link ID. As neighboring nodes establish new links, link IDs are communicated during the link establishment process. All data transmissions between the two nodes use the same link.

Mesh mode uses two types of scheduling, distributed and centralized. In distributed scheduling, all the nodes must coordinate their transmissions in their extended neighborhood. This can be accomplished by having every node broadcasting its schedule (available resources, requests, and grants) to all its neighbors. Schedules may also be established by directed uncoordinated requests and grants between two nodes. Before transmitting, a node must ensure that it will not cause collisions with the transmissions scheduled by any other node in its extended neighborhood.

In centralized scheduling, resource request from all the mesh subscriber stations within a certain hop range is gathered by the mesh base station. The base station determines the amount of resources it wishes to grant on each link in the network, and communicates the grants to all the mesh subscriber stations in the hop range.

11.2.2.4 QoS

WiMAX was designed with QoS in mind, to provide low latency for delay-sensitive services and data prioritization. QoS support resides within the MAC layers of the base station and subscriber stations. The base station contains a packet queue for each downlink connection. It uses the QoS parameters and the status of the queues to determine which queue to use for the next SDUs to be sent. The subscriber station has similar queues for uplink connections [6].

Bandwidth is granted to the subscriber stations from the base stations when it is needed. Subscriber stations can request bandwidth in a few different ways. Using unsolicited granting, during the setup of an uplink connection, subscriber stations request a fixed amount of bandwidth on a periodic basis. Once the connection is complete the subscriber stations cannot request any more bandwidth. The base station can use broadcast polls to determine if subscriber stations need bandwidth. An issue arises when two or more stations respond to the same poll causing a collision. After collision, nodes follow an exponential backoff algorithm and wait to respond again. Bandwidth requests can also be piggybacked on a PDU sent from the subscriber station.

The base station's uplink scheduler uses the bandwidth requests to estimate the remaining backlog at each uplink connection. It uses this knowledge and the set of QoS parameters to determine future uplink grants. While the bandwidth requests are per connection, the base station grants uplink capacity to each subscriber station as a whole. Therefore, the subscriber station also implements a scheduler within its MAC to allocate its uplink bandwidth between its connections.

11.3 Security Features

WiMAX security has two goals, one is to provide privacy across the wireless network and the other is to provide access control to the network. Privacy is accomplished by encrypting connections between the subscriber station and the base station. The base station protects against unauthorized access by enforcing encryption of service flows across the network. A privacy and key management (PKM) protocol is used by the base station to control the distribution of keying data to subscriber stations. This allows the subscriber and base stations to synchronize keying data. Digital-certificate-based subscriber station authentication is included in the PKM to provide access control [5].

11.3.1 Security Associations

A security association (SA) is the set of security information a base station and one or more of its client subscriber stations share to support secure communication across a WiMAX network. WiMAX uses two different types of SAs, data and authorization [5,7].

There are three different types of data SAs: primary, static, and dynamic. Primary SAs are established by the subscriber stations during their initialization process. The base station provides the static SAs. Dynamic SAs are established and eliminated as needed for service flows. Both static and dynamic SAs can be shared among multiple subscriber stations [5].

TABLE 11.4

Contents of Data SAs

16-bit SA identifier (SAID)
Encryption cipher to protect the data exchanged over the connection
Two TEKs: one for current operation and another for when the current key expires
Two 2-bit key identifiers, one for each TEK
TEK lifetime. The minimum value is 30 min and the maximum value is 7 days. The default is half a day
Initialization vector for each TEK
Data SA type indicator (primary, static, dynamic)

Table 11.4 shows the contents of a data SA. The SA identifier (SAID) is used to uniquely identify the data SA. The encryption cipher defines what method of encryption will be used to encrypt data. Initially, the IEEE 802.16 standard defined the use of the data encryption standard (DES) in cipher block chaining (CBC) mode. Later, in the IEEE 802.16e revision, more modes were defined. Section 11.3.4 covers data encryption in detail.

Traffic encryption keys (TEKs) are used to encrypt data transmissions between the base stations and subscriber stations. The data SA defines two TEKs, one for current operations and a second to be used when the current one expires. Two TEK identifiers are included, one for each key. A TEK lifetime is also included to indicate when the TEK expires. The default lifetime is half a day, but it can vary from 30 min to 7 days.

DES in CBC mode requires an initialization vector to operate. Therefore, one for each TEK is included in the data SA. Both initialization vectors are 64 bits in length to accommodate the 64-bit block size used in DES encryption.

The data SA type is also included to indicate whether it is a primary, static, or dynamic data SA.

Data SAs protect transport connections between one or more subscriber stations and a base station. Subscriber stations typically have one SA for their secondary management channel and either one SA for both uplink and downlink transport connections or separate SAs for uplink and downlink connections. For multicasting, each group requires an SA to be shared among its members; therefore the standard lets many connection IDs share a single SA [7].

Authorization SAs are shared between a base station and a subscriber station. They are used by the base station to configure data SAs for the subscriber station [7].

Table 11.5 shows the contents of an authorization SA. An X.509 certificate is included, which allows the base station to identify the subscriber station. Section 11.3.2.2 goes into detail about X.509 certificates and how they are used.

The 160-bit authorization key (AK) is included to allow the base station and subscriber station to authenticate each other during TEK exchanges. Section 11.3.3.2 describes the TEK exchange process. A 4-bit AK identifier is used to distinguish among different AKs. An AK lifetime is also included

TABLE 11.5

Contents of Authorization SAs

X.509 certificate identifying the subscriber station
160-bit authorization key
4-bit authorization key identifier
Authorization key lifetime. The minimum value is 1 day and the value maximum is 70 days. The default is 7 days
Key encryption key (KEK) for distributing TEKs
Downlink hash function-base message authentication code (HMAC) key
Uplink HMAC keys
List of authorized data SAs

to indicate when the AK expires. The default lifetime is 7 days, but it can range from 1 to 70 days.

Key encryption keys (KEKs) are used to encrypt TEKs during the TEK exchange process. Two KEKs are required for the encryption process and are derived from the AK. The KEKs are computed by first concatenating the hex value 0x53 repeated 64 times and the AK. Then the SHA-1 hash of this value is computed, which outputs 160 bits. Finally, the first 128 bits of the output are taken and divided into two 64-bit TEKs. These two TEKs are included in the authorization SA.

Two hashed message authentication code (HMAC) keys, one for uplink and one for downlink, are included to allow for the creation of HMACs during the TEK exchange process. The uplink key is used to create an HMAC of messages to be sent, while the downlink key is used to create an HMAC of messages received, allowing the receiver to authenticate the message. The uplink key is obtained by concatenating the hex value 0x3A repeated 64 times and the AK, then computing the SHA-1 hash of this value, creating a 160-bit HMAC key. The downlink key is computed in the same fashion, but the hex value 0x5C is concatenated with the AK instead.

A list of authorized data SAs is also included in the authorization SA that provides the subscriber station with the knowledge of the data SAs it can request.

11.3.2   Authentication

11.3.2.1   *Hashed Message Authentication Code*

HMACs are used to provide message authentication. By using HMACs, the receiver can verify who sent the message. This is possible because the sender creates an HMAC of the message it wishes to send using a key known only by the sender and receiver. When the receiver gets the message, it computes its own HMAC of the message using the same key and compares the one it computed with the one received from the sender. If the HMACs match then the sender is confirmed.

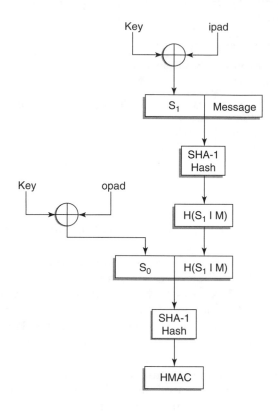

FIGURE 11.5
HMAC creation.

HMACs are created as a function of a key and the message. Figure 11.5 illustrates the HMAC creation process. First, the hash key, defined in the authorization SA, is exclusive-ored (XORed) with an ipad, which is the byte 0x36 repeated 20 times to match the size of the hash key. This 160-bit value is appended to the beginning of the message, which is then hashed. The IEEE 802.16 standard defines the use of SHA-1 to compute the hash.

The hash key is then XORed with an opad, which is the byte 0x5C repeated 20 times to match the size of the hash key. This 160-bit value is appended to the beginning of the output of the previous hash. These two values are then hashed to produce the HMAC.

11.3.2.2   X.509 Certificates

X.509 certificates are used to allow the base station to identify subscriber stations. Table 11.6 describes the required fields as defined by the IEEE 802.16 standard. While extension data may be included, the standard does not define any [5,7].

There are two types of certificates: manufacturer certificates and subscriber station certificates. A manufacturer certificate, which identifies the

TABLE 11.6

X.509 Certificate Fields

X.509 Certificate Fields	Description
Version	Indicates the X.509 certificate version
Serial number	Unique integer assigned by the issuing CA
Signature	Object identifier and optional parameters defining algorithm used to sign the certificate
Issuer	Name of CA that issued the certificate
Validity	Period for which certificate is valid
Subject	Name of entity whose public key is certified in the subject public key info field
Subject public key info	Contains the public key, parameters, and the identifier of the algorithm used with the key
Issuer's unique ID	Optional field to allow reuse of issuer names over time
Subject's unique ID	Optional field to allow reuse of subject names over time
Extensions	The extension data
Signature algorithm	Object identifier and optional parameters defining algorithm used to sign the certificate
Signature value	Digital signature of the abstract syntax notation 1 distinguished encoding rules encoding of the rest of the certificate

manufacturer of the device, can be a self-signed certificate or issued by a third party. A subscriber station certificate is typically created and signed by the manufacturer of the station. It is used to identify a subscriber station and includes the MAC address of the station in the subject field. Base stations can use the manufacturer certificate to verify the subscriber station's certificate, allowing it to determine if the device is legitimate [7].

11.3.2.3 *Extensible Authentication Protocol*

The IEEE 802.16e standard introduced an alternative to the authentication scheme based on X.509 certificates. This new scheme is considered to be more flexible and is based on the extensible authentication protocol (EAP) [7].

To obtain authentication during link establishment, EAP messages are encoded directly into management frames. Two additional PKM messages, PKM EAP request and PKM EAP response, were added to transport EAP data.

Currently, EAP methods to support the security needs of wireless networks is an active area of research and, therefore, the IEEE 802.16e standard does not specify a particular EAP-based authentication method to be used.

11.3.3 Privacy and Key Management

Subscriber stations use the PKM protocol to obtain authorization and traffic keying material from the base station. The PKM protocol can be broken into two parts. The first handles subscriber station authorization and AK exchange. The second handles TEK exchange [5].

11.3.3.1 Authorization and AK Exchange

PKM authorization is used to exchange an AK from the base station to the subscriber station. Once the subscriber station receives an initial authorization, it will periodically seek reauthorization. The AK exchange is accomplished using three messages, illustrated in Figure 11.6 [5,7].

The subscriber station initiates the exchange by sending a message containing the subscriber station manufacturer's X.509 certificate to the base station. The message is strictly informative and can be ignored by the base station. However, base stations can be configured to only allow access to devices from trusted manufacturers.

The second message is sent from the subscriber station to the base station immediately after the first message. This message is a request for an AK and a list of SAIDs that identify SAs the subscriber station is authorized to participate in. There are three parts to the message: a manufacturer-issued X.509 certificate, cryptographic algorithms supported by the subscriber station, and the SAID of its primary SA.

The base station uses the subscriber station's certification to determine if it is authorized. If it is, the base station will respond with the third message. The base station uses the subscriber station's public key, obtained from its certification, to encrypt the AK using RSA. The encrypted AK is then included in the message along with the SeqNo, which distinguishes between successive AKs, the key lifetime, and a list of SAIDs of the static SAs the subscriber station is authorized to participate in.

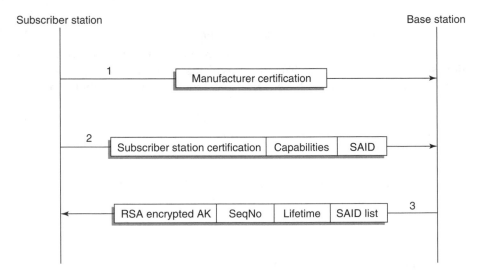

FIGURE 11.6
PKM authorization.

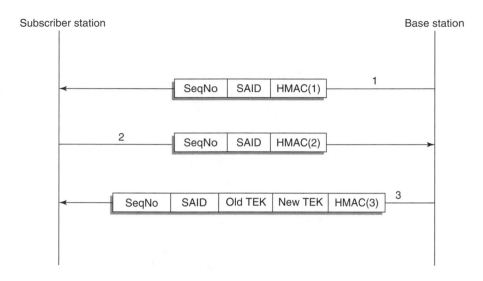

FIGURE 11.7
PKM TEK exchange.

11.3.3.2   TEK Exchange

Once the subscriber station has been authorized, it will establish an SA for each SAID in the list received from the base station. This is accomplished by initiating a TEK exchange. Once an SA is established, the subscriber station will periodically refresh keying material. The base station can also force rekeying if needed. Figure 11.7 illustrates the TEK exchange process [5,7].

The first message of a TEK exchange is optional and allows the base station to force rekeying. There are three parts to the message: SeqNo refers to the AK used in creating the HMAC, the SAID refers to the SA that is being rekeyed, and the HMAC allows the subscriber station to authenticate the message.

The second message is sent by the subscriber station in response to the first message or if the subscriber station wants to refresh the keying material. There are three parts to the message: SeqNo refers to the AK used in creating the HMAC, the SAID refers to either the SAID received in the first message or one of the SAs from the subscriber station's authorized SAID list, and the HMAC allows the base station to authenticate the message.

If the HMAC in the second message is valid then the base station will send the third message. As in the first two messages, a SeqNo, the SAID, and the HMAC are included. In addition to these the old TEK and a new TEK are added. The old TEK just reiterates the active SA parameters while the new TEK is to be used when the active one expires. The base station encrypts both the old and new TEKs using triple DES in electronic code book (ECB) mode with the KEK associated with the SA.

Figure 11.8 illustrates the TEK encryption process. Section 11.3.1 describes how the KEK is created. Here, KEK 1 is the leftmost 64 bits of the computed

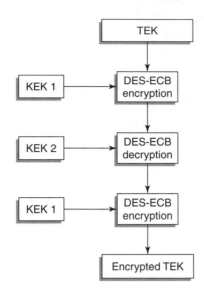

FIGURE 11.8
TEK encryption process.

KEK and KEK 2 is the rightmost 64 bits. These two keys are used in the triple DES encryption in which the TEK is first encrypted using KEK 1. The output is then decrypted using KEK 2 and then encrypted using KEK 1. This process is performed on both the old and new TEKs to produce two encrypted TEKs.

11.3.4   Data Encryption

To provide privacy for the data being transmitted in WiMAX networks, the IEEE 802.16 standard employed the use of DES in CBC mode. Currently, DES is considered to be insecure and has been replaced by the AES. Therefore, the IEEE 802.16e standard defines the use of AES for use in encryption [7].

11.3.4.1   DES

Using DES in CBC mode, the payload field of the MAC PDU is encrypted, but the GMH and CRC are not. Figure 11.9 illustrates the encryption process.

    CBC mode requires an initialization vector (IV), which is computed by taking the XOR of the IV parameter in the SA and the content of the PHY synchronization field. The DES encryption process uses the IV and the TEK from the SA of the connection to encrypt the payload of the PDU. This ciphertext payload then replaces the original plaintext payload. The EC bit in the GMH will be set to 1 to indicate an encrypted payload and the EKS bits will be set to indicate that the TEK was used to encrypt the payload. If the CRC is included, it will be updated for the new ciphertext payload [5].

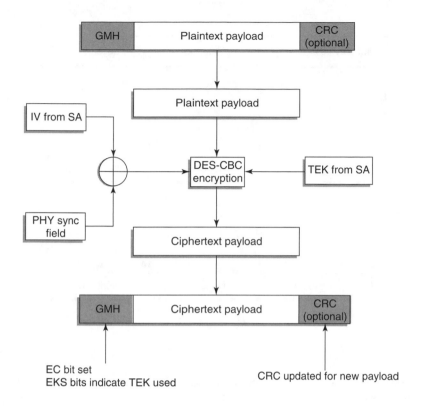

FIGURE 11.9
DES-CBC encryption.

11.3.4.2   AES

The IEEE 802.16e standard added the use of AES to provide stronger encryption of data. It defines the use of AES in four modes: CBC, counter encryption (CTR), CTR with CBC message authentication code (CCM), and ECB. CTR mode is considered better than CBC mode due to its ability to perform parallel processing of data, preprocessing of encryption blocks, and is simpler to implement. CCM mode adds the ability to determine the authenticity of an encrypted message to CTR mode. ECB mode is used to encrypt TEKs.

11.3.4.2.1   AES in CCM Mode

AES-CCM requires that the transmitter constructs a unique nonce, which is a per-packet encryption randomizer. IEEE 802.16e defines a 13-byte nonce, as shown in Figure 11.10. Bytes 0–4 are constructed from the first 5 bytes of the GMH. Bytes 5–8 are reserved and are all set to 0. Bytes 9–12 are set to the packet number (PN). The PN is associated with an SA and set to 1 when the SA is established and when a new TEK is installed. Since the nonce is dependent on the GMH, modifications to the GMH can be detected by the receiver [7,8].

0   4 5   8 9   12

| First 5 bytes of GMH | 0x00000000 | Packet number |

FIGURE 11.10
CCM nonce.

0   1   13 14   15

| Flag (0x19) | Nonce | Length of payload |

FIGURE 11.11
CCM CBC block.

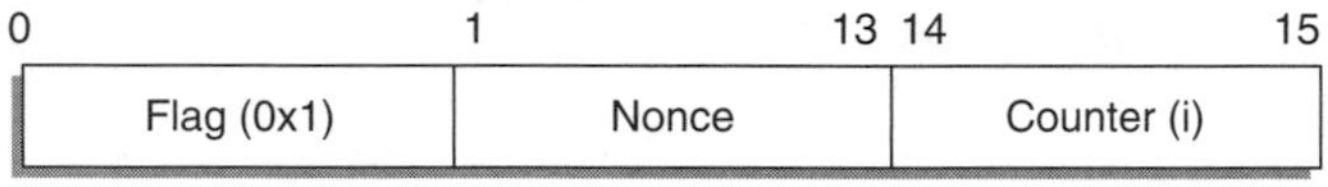

0   1   13 14   15

| Flag (0x1) | Nonce | Counter (i) |

FIGURE 11.12
CCM counter block.

To create a message authentication code, AES-CCM uses a variation of CBC mode. Instead of using an IV, an initial CBC block is appended to the beginning of the message before it is encrypted. As seen in Figure 11.11, the initial CBC block consists of a flag, the packet nonce, and the length of the payload.

To encrypt the payload and the message authentication code, AES-CCM uses CTR mode. With this mode, n counter blocks are created, where n is the number of blocks needed to match the size of the message plus one block for the message authentication code (AES uses 128-bit block sizes). The first block is used for encrypting the message authentication code and the remaining blocks are used to encrypt the payload. As seen in Figure 11.12, the counter block consists of a flag, the packet nonce, and the block number i, where i goes from 0 to n.

The message authentication code is created by encrypting the initial CBC block and plaintext payload. Figure 11.13 illustrates the message authentication code creation and subsequent encryption of the message authentication code.

The first step in creating the message authentication code is to extract the plaintext payload from the PDU and append the initial CBC block to the beginning of it. This is then encrypted using AES in CBC mode with the TEK from the SA of the connection. The last 128 bits (size of one AES block) of the encrypted output is selected to represent the message authentication code.

The sender will perform this process and then encrypt the message authentication code with the message. The receiver will decrypt the message and message authentication code and then perform the same process on

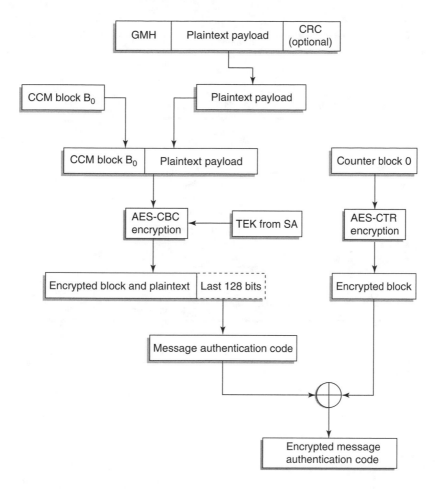

FIGURE 11.13
AES-CCM message authentication creation and encryption.

the message. The receiver will then compare the message authentication code it created with the one received. If they are the same, the message is authenticated, if not the message is discarded.

Encryption of the message authentication code is accomplished by encrypting counter block 0 using AES in CTR mode with the TEK from the SA of the connection. This encrypted block is then XORed with the message authentication code to produce the encrypted version.

Payload encryption is accomplished by first encrypting counter blocks 1 through n with AES in CTR mode using the same TEK used to encrypt the message authentication code. The plaintext payload is then extracted from the PDU and XORed with the encrypted counter blocks. This produces the ciphertext payload, as shown in Figure 11.14.

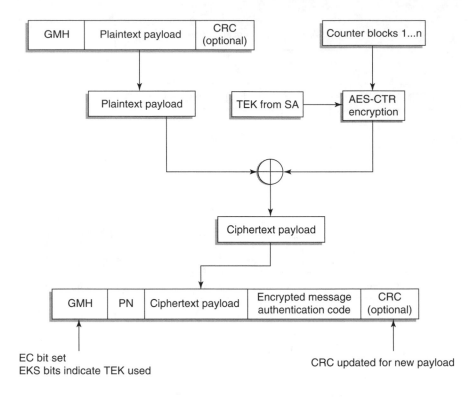

FIGURE 11.14
AES-CCM payload encryption.

The PN is then appended before the ciphertext payload and the message authentication code is appended after the ciphertext payload. This set of data then replaces the plaintext payload. The EC bit in the GMH will be set to 1 to indicate an encrypted payload and the EKS bits will be set to indicate the TEK used to encrypt the payload. If the CRC is included, it will be updated for the new payload.

11.4   Open Issues

In WiMAX, security threats apply to both the PHY and MAC layers. Possible PHY level attacks include jamming of a radio spectrum, causing denial of service to all stations, and flooding a station with frames to drain its battery. Currently, there are no efficient techniques available to prevent PHY layer attacks. Therefore, the focus of WiMAX security is completely at the MAC level [7]. In this section, we discuss some of the open security issues in the WiMAX networks.

11.4.1 Authorization Vulnerabilities

A major vulnerability of WiMAX security is the lack of a base station certificate, which is needed for mutual authentication. Without mutual authentication, the subscriber stations cannot verify that authorization protocol messages received are from the base station. This leaves the subscriber station open to forgery attacks, allowing any rogue base station to send it responses [7].

A solution to issues with WiMAX's authentication and authorization procedures is proposed in Ref. 9. It proposes the wireless key management infrastructure (WKMI), which is based on the IEEE 802.11i standard. WKMI is a key management hierarchy infrastructure that is based on the use of X.509 certificates allowing subscriber stations and base stations to perform mutual authentication and key negotiation.

AK generation is another concern with the authorization protocol. Though the standard assumes a random AK generation, it imposes no requirements. An additional weakness lies in the fact that the base stations generate the AK, requiring the subscriber station to trust that the base station always generates a new AK that is cryptographically separated from all other AKs previously generated. To hold true, the base stations must have a perfect random number generator. Allowing both the subscriber station and base station to contribute to the AK generation could solve this issue [7].

11.4.2 Key Management

A major issue with key management in WiMAX is the size of its TEK identifier. Currently, a 2-bit number is used, which allows only four values (0 to 3) to be represented. This causes the TEK identifier to wrap from 3 to 0 on every fourth key, leaving stations open to replay attacks in which an attacker could reuse expired keys. To solve this issue, the TEK identifier's size needs to be increased to prevent wrapping. If the longest AK lifetime (70 days) and the shortest TEK lifetime (30 min) are considered, then 3360 different TEKs need to be represented, which would require 12 bits be used for the TEK identifier [7].

Another issue is the TEK lifetime, which can be set anywhere between 30 min and 7 days with a default of half a day. If DES in CBC mode is used for encryption with the possible lifetime values, the security of the data may be compromised. This is due to the fact that DES in CBC mode becomes insecure after operating on $2^{n/2}$ blocks with the same encryption key, where n is the block size. Since DES uses a 64-bit block size, after 2^{32} blocks the encryption will be insecure. The time it takes to happen depends on the average throughput between stations. Considering the high transfer rates WiMAX offers and the ability to choose a larger TEK lifetime, encryption insecurity is highly possible.

The introduction of AES in the IEEE 802.16e standard will help solve the TEK lifetime issues. Unfortunately, implementation of this standard is still a way off, possibly leaving current deployments of WiMAX insecure.

References

1. Z. Abichar, P. Yanlin, and J.M. Chang, WiMAX: The emergence of wireless broadband, *IT Professional*, Vol. 8, pp. 44–48, 2006.
2. J.P. Conti, The long road to WiMAX [wireless MAN standard], *IEE Review*, Vol. 51, pp. 38–42, 2005.
3. B. Rathgeb and C. Qiang, *Utilizing the IEEE 802.16 Standard for Homeland Security Applications*, Orlando, FL, 2005.
4. M. Donahoo and B. Steckler, *Emergency Mobile Wireless Networks*, Atlantic City, NJ, 2005.
5. *IEEE standard for local and metropolitan area networks—Part 16: Air interface for fixed broadband wireless access systems*, IEEE Std 802.16-2004 (Revision of IEEE Std 802.16-2001), pp. 851–857, 2004.
6. C. Cicconetti, L. Lenzini, E. Mingozzi, and C. Eklund, Quality of service support in IEEE 802.16 networks, *Network, IEEE*, Vol. 20, pp. 50–55, 2006.
7. D. Johnston and J. Walker, Overview of IEEE 802.16 security, *IEEE Security & Privacy*, Vol. 2, pp. 40–48, 2004.
8. *IEEE standard for local and metropolitan area networks—Part 16: air interface for fixed and mobile broadband wireless access systems amendment 2: Physical and medium access control layers for combined fixed and mobile operation in licensed bands and corrigendum 1*, IEEE Std 802.16e-2005 and IEEE Std 802.16-2004/Cor 1-2005 (Amendment and Corrigendum to IEEE Std 802.16-2004), pp. 801–822, 2006.
9. Y. Fan, Z. Huaibei, Z. Lan, and F. Jin, *An Improved Security Scheme in WMAN Based on IEEE Standard 802.16*, Wuhan, China, 2005.

12

WiMAX Security: Privacy Key Management

Nirwan Ansari, Chao Zhang, Yuanqiu Luo, and Edwin Hou

CONTENTS

12.1   WiMAX Overview

WiMAX stands for worldwide interoperability for microwave access. It was proposed to facilitate high-speed data distribution through wireless metropolitan area networks (WMANs). With the advantages of rapid deployment, high scalability, and low upgrade cost, WiMAX attempts to tackle the last mile bottleneck problem of current telecommunications networks.

The IEEE 802.16 working group on broadband wireless access (BWA) standards develops standards and recommends practices to support the development and deployment of the WiMAX technology.

The first WiMAX standard, i.e., IEEE 802.16-2001 [1], was published in 2002. It defines a point-to-multipoint (PMP) fixed wireless access system between a base station (BS) and its associated subscriber stations (SSs). IEEE 802.16-2001 operates in the 10–66 GHz frequency range, which is the so-called line-of-sight (LOS) communications. The IEEE 802.16-2004 standard [2] was published in 2004 to extend the WiMAX specification into the 2–11 GHz frequency range, the so-called nonline-of-sight (NLOS) operation. IEEE 802.16-2004 also describes the WiMAX system profiles and conformance criteria to adapt to the dynamic wireless environment. By introducing the mesh mode, IEEE 802.16-2004 is capable of forwarding traffic from a node to its neighboring nodes. The latest WiMAX standard, IEEE 802.16e-2005 [3], was approved in December 2005. By employing scalable orthogonal frequency division multiplexing (SOFDM), IEEE 802.16e-2005 provides full mobility support for both licensed and unlicensed spectra. The aforementioned WiMAX standards herald a promising new tool for broadband access in the effort to bridge the bandwidth mismatch and to support user mobility.

As illustrated in Figure 12.1, WiMAX standards define the protocol structure at both the medium access control (MAC) layer and the physical (PHY) layer. The WiMAX PHY layer supports flexible operation across a wide range of spectrum allocations (from 2 to 66 GHz), including variations in channel bandwidth, frequency division duplex, and time division duplex. The WiMAX MAC layer is defined to provide a common feature set across diverse PHY performance. The major MAC functionalities cover initial ranging, network entry, bandwidth requests, connection-oriented management, as well as information security through the dynamic WiMAX environment.

Communications in WiMAX are connection-oriented. All services from the upper protocol layer above WiMAX MAC, including the connectionless services, are mapped into connections between the SS and the BS in the WiMAX MAC layer. One SS may have multiple connections to its associated BS with the purpose to provide diverse services to the subscribers. Connections are identified by 16-bit connection identifiers (CIDs). Such a connection-based mechanism facilitates bandwidth arbitration and QoS support in the dynamic wireless environment. The WiMAX MAC layer is thus defined to support the connection-oriented service in an organic manner.

Among the three sublayers in WiMAX MAC, the service-specific convergence sublayer (CS) connects the MAC layer with the upper layer. After classifying service data units (SDUs) from upper layer protocols, the CS sublayer associates the SDUs to the proper MAC service flow identifier (SFID) and CID. For different upper layer protocols, such as ATM, Ethernet, and IP, the CS sublayer defines different specifications accordingly. Therefore, the MAC common part sublayer (CPS) does not need to understand the format of or parse any information from the CS payload. The CPS sublayer of the

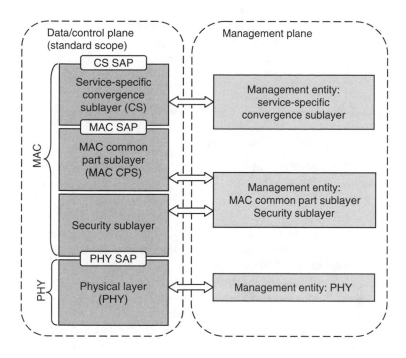

FIGURE 12.1
WiMAX standard protocol structure. (Adapted from IEEE Std. 8802.16e-2005, "IEEE Standard for Local and Metropolitan Area Networks—Part 16: Air Interface for Fixed and Mobile Broadband Wireless Access Systems," IEEE, 2006.)

WiMAX MAC is responsible for providing the functionalities, including system access, bandwidth allocation, and WiMAX connection establishment and maintenance. It exchanges MAC SDUs (MSDUs) with various CSs.

The security sublayer plays a key role in authentication, key establishment, as well as information encryption. It exchanges MAC protocol data units (MPDUs) directly with PHY. Toward the end of handling the dynamic wireless environment, WiMAX specifies a set of privacy and key management mechanisms. The two components in the security sublayer are the encapsulation protocol and privacy key management (PKM) protocols. The encapsulation protocol encrypts WiMAX data across BWA, while the PKM protocols ensure the secure distribution of keying material and authorized access to the connections between the SS and the BS. As a safeguard to high-speed broadband access with flexible mobility, the WiMAX security sublayer provides the SS with privacy and protects the BS from malicious attacks.

This chapter presents an overview of the WiMAX security mechanism, particularly, the management schemes for subscriber privacy and server security. Section 12.2 describes the challenges to WiMAX security. Section 12.3 presents PKM version 1 (PKMv1), the fundamental security mechanism for

WiMAX communications. Section 12.4 elaborates PKM version 2 (PKMv2), an enhanced security mechanism with a major improvement for mutual authentication. Section 12.5 concludes this chapter with a discussion on the open issues in WiMAX security.

12.2 WiMAX Security Challenges

As WiMAX standards expand from considering a fixed LOS and PMP high-frequency system (10–66 GHz) to including a lower frequency (2–11 GHz) NLOS mobile system, WiMAX is open to more security threats than other wireless systems. Attacks against the original standard, IEEE 802.16-2001, require an adversary to physically place the attacking equipment between the SS and the BS, and the equipment has to be able to operate at the comparatively high frequencies of 10–66 GHz. The IEEE 802.16-2004 standard defines operations at lower frequencies, thus reducing the hardware implementation complexity and the physical placement constraints. As a result, new security challenges emerge especially for the mesh mode, such as the trustworthiness of the next-hop mesh node. The IEEE 802.16e-2005 standard accommodates user mobility, hence facilitating attackers to easily stage an attack. With less constraint on the physical location, the management messages become more vulnerable to attackers. Since WiMAX uses air interface for the transmission medium, both the PHY and MAC layers are readily exposed to security threats [4,5].

12.2.1 Physical Layer Threats

Two principal threats to the WiMAX PHY are jamming and scrambling [5]. Jamming is achieved by introducing a source of noise strong enough to significantly reduce the capacity of the WiMAX channel. The information and equipment required to perform jamming are not difficult to acquire. Resilience to jamming can be augmented by increasing the power of signals or increasing the bandwidth of signals via spreading techniques such as frequency hopping or direct sequence spread spectrum. The practical options include a more powerful WiMAX transmitter, a high gain WiMAX transmission antenna, or a high gain WiMAX receiving antenna. It is easy to detect jamming in WiMAX communications as it can be heard by the receiving equipment. Law enforcement can also be involved to stop jammers. Since jamming is fairly easy to detect and address, we believe that it does not pose a significant impact on both the WiMAX users and systems.

Scrambling is usually instigated for short intervals of time and is targeted to specific WiMAX frames or parts of frames. WiMAX scramblers can selectively scramble control or management messages with the aim of affecting the normal operation of the network. Slots of data traffic belonging to the targeted SSs can be scrambled selectively, forcing them to retransmit.

The attacker, often behaved as a WiMAX SS, can reduce the effective bandwidth of the victims, i.e., other SSs, and accelerate the processing of its own data by selectively scrambling the uplink slots of other SSs. Unlike the random behavior of a WiMAX jammer, a scrambler needs to interpret WiMAX control information correctly and to generate noise during specific intervals. Hence, attacks from scrambling are intermittent, and thus exacerbate the detection task. Monitoring anomalies beyond the performance norm is a viable means to detect scrambling and scramblers.

12.2.2 MAC Layer Threats

MPDU is the data unit transmitted in the WiMAX MAC layer. As shown in Figure 12.2, MPDU uses different formats to carry different information. The common format of each MPDU consists of a MAC header, service data, and an optional cyclic redundancy check (CRC). The unencrypted generic MAC header format contains the specific encryption information in the MAC header. Encryption is applied to the MAC PDU payload.

All MAC management messages shall be sent unencrypted to facilitate registration, ranging, and normal operation of the MAC. The WiMAX management messages are carried in the MPDU as illustrated in Figure 12.2b. WiMAX encrypts neither the MAC headers nor the MAC management messages, with the purpose to enable various operations of the MAC layer. Therefore, an attacker, as a passive listener of the WiMAX channel, can retrieve valuable information from unencrypted MAC management messages. Eavesdropping of management messages may reveal network topology to the eavesdropper, posing a critical threat to SSs as well as the WiMAX system. WiMAX requires device-level authentication to tackle this problem. The main idea is to issue a WiMAX device with an Rivest-Shamir-Adleman (RSA)/X.509 digital certificate from the manufacturer. The digital certificate is employed for authentication and authority detection. The unauthenticated device is blocked from eavesdropping of the network.

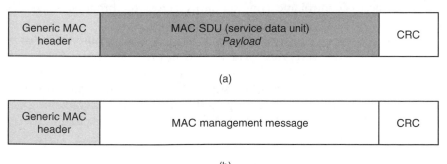

FIGURE 12.2
MAC protocol data unit structure. (a) Transport connection MPDU. (b) Management connection MPDU.

Identity theft is a severe threat to unlicensed services supported by WiMAX [5,6]. A fake device can use the hardware address of another registered device by intercepting management messages over the air. Once succeeded, an attacker can turn a BS into a rogue BS. A rogue BS can imitate a legitimate BS by confusing the associated SSs. Those SSs try to acquire WiMAX services from the rogue BS, resulting in degraded service or even service termination.

The Wireless-fidelity (Wi-Fi) network employs carrier sense multiple access (CSMA), and thus identity theft has become one of the top security threats. The reason is that the attacker can easily capture the identity of a legitimate access point (AP) by listening to the CSMA process, which readily reveals information on the AP identity. The attacker can then construct a message by using the legitimate AP's identity, wait until the medium is idle, and distribute the malicious message.

In WiMAX, time division multiple access (TDMA) is adopted. To steal the identity, the attacker must transmit while the legitimate BS is transmitting, and the signal of the attacker must arrive at the targeted SSs with high enough strength to subside the signal of the legitimate BS in the background. Since the transmission is divided into time slots, the attacker has to interpret the time slot allocated to the legitimate BS successfully and detect the BS signal strength correctly, both of which make identity theft more difficult and challenging. Besides, mutual authentication has been introduced into the latest WiMAX standard, further reducing the likelihood of identity theft. In the following sections, we will elaborate the PKM protocols, the security management mechanism to effectively overcome identity theft and eavesdropping in WiMAX.

12.3   Privacy Key Management Protocol Version 1

The security sublayer is defined at the bottom of the WiMAX MAC layer to provide access control and confidentiality across the broadband wireless network through encryption and key management.

Figure 12.3 illustrates the protocol stack of the security components of the WiMAX system. The PKM protocol in the middle provides secure distribution of keying data from the BS to the SS. PKM manages the key exchange process and the procedure for applying the supported encryption and authentication algorithms to MPDUs. By specifying the synchronization of keying data between the BS and SS, PKM enforces the conditional access to a particular WiMAX connection.

12.3.1   Security Procedure

WiMAX communications follow the security procedure defined in PKMv1 to ensure secure access of a connection. As shown in Figure 12.4, authentication

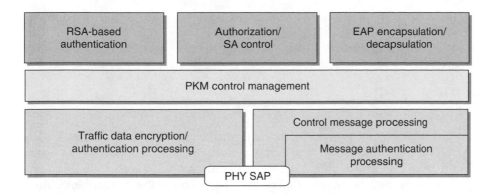

FIGURE 12.3
Protocol stack of the security sublayer. (Adapted from IEEE Std. 8802.16e-2005, "IEEE Standard for Local and Metropolitan Area Networks—Part 16: Air Interface for Fixed and Mobile Broadband Wireless Access Systems," IEEE, 2006.)

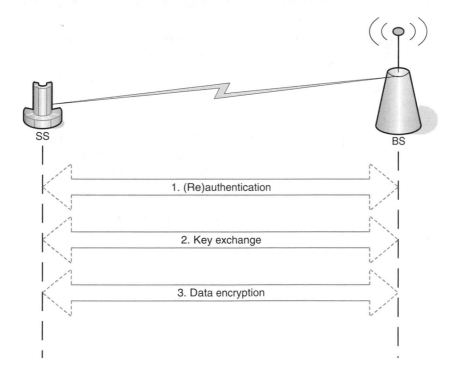

FIGURE 12.4
WiMAX security procedure.

is conducted as the first step of security enforcement prior to any data transmission. When an SS enters the WiMAX network, the BS verifies the SS identity, followed by the key exchange step. Once the SS identity is authenticated and a key is successfully established, the BS registers the SS into the network and the key is used to encrypt the data transmitted through the WiMAX connection.

The remaining part of Section 12.3 will elaborate each step of the security procedure.

12.3.2 Authentication

Authorization is the process for authenticating a client SS's identity by the BS. An SS starts authorization by sending an authentication information message to the target BS, containing the SS manufacturer's X.509 certificate [7] issued by the manufacturer or an external authority. Following the authentication information message, an authorization request message is sent immediately to the BS to request for an authentication key, with the following information from the SS for security authentication:

- The manufacturer-issued X.509 certificate (the requesting SS's identification)
- A description of the cryptographic algorithms that the requesting SS supports (the so-called security association [SA])
- The SS's basic CID, which is equal to its primary security association identifier (SAID)

The detailed process of security authentication is shown in Figure 12.5. In the authentication process, WiMAX standards define the term "security association" to specify the set of security information a BS and its SS (or SSs) share. SA, identified with a SAID, is essentially the set of security information a BS and its SSs support for secure communications. It includes the cryptographic suites and keys for encryption. As illustrated in Figure 12.5, an SS informs the BS of its SAID. The BS validates the requesting SS's identity by determining the encryption algorithms and protocols it shares with the SS. The BS also determines whether the SS is authorized for basic unicast services and any other services provided by the WiMAX network.

After verifying the requesting SS's identity, the BS activates an authentication key (AK) for the SS, encrypts it with the SS's public key, and sends it back to the SS in an authorization reply message. Authorization reply includes the AK encrypted with the SS's public key, a 4-bit key sequence number (used to distinguish between successive AKs), a key lifetime, and the identities and properties of the SA list the SS has been authorized to access.

With the authentication process, the BS associates the SS's authenticated identity to a paying subscriber, and hence to the data services that the subscriber is authorized to access. With the AK exchange, the BS determines the authenticated identity of the client SS and the services the SS is authorized to access. Since the BS authenticates the SS, it protects against an attacker from employing a cloned SS, masquerading as a legitimate subscriber's SS.

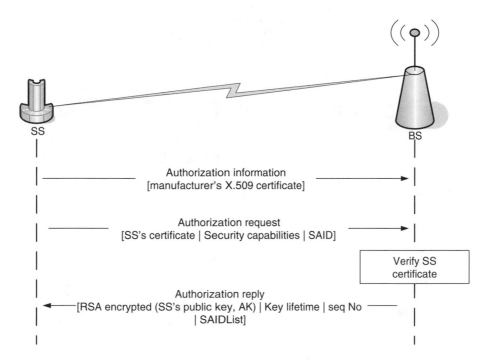

FIGURE 12.5
The authentication process.

PKMv1 mandates the use of X.509 digital certificates together with the RSA public-key encryption algorithm [8] to conduct authentication; readers are referred to Appendix III for details of the RSA.

12.3.3 Key Exchange

There are five kinds of keys used to secure WiMAX communications: AK, key encryption key (KEK), downlink hash function-based message authentication code (HMAC) key, uplink HMAC key, and traffic encryption key (TEK). AK is activated by a BS during the authentication process. As the shared secret between the SS and the BS, AK is used to secure subsequent key exchanges in PKMv1.

As shown in Figure 12.6, a 128-bit AK is used to generate the 128-bit KEK by the BS. KEK is used for TEK encryption and distribution. The KEK is derived from the AK by the following formula:

$$KEK = \text{Truncate\_128}\{SHA1[(AK \mid 0^{44}) \oplus 53^{64}]\} \qquad (12.1)$$

In Equation 12.1, AK concatenates with 0^{44} and XORs 53^{64} as denoted by $(AK \mid 0^{44}) \oplus 53^{64}$. The result is hashed by the secure hash algorithm SHA1, the most commonly used hash function defined by the secure hash standard.

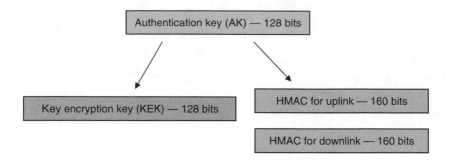

FIGURE 12.6
The key derivation process.

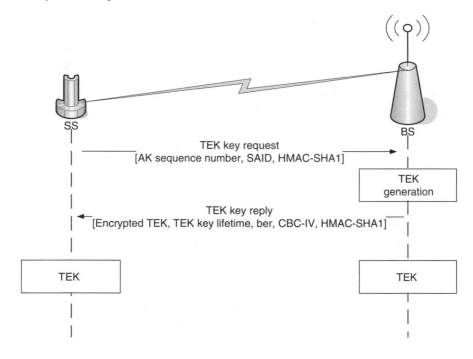

FIGURE 12.7
The TEK exchange procedure.

Truncate_128(.) retrieves the first 128 bits of the hash result as the KEK and discards the rest of the bits.

The downlink HMAC key and uplink HMAC key provide data authenticity of key distribution messages from the BS to the SS and from the SS to the BS, respectively. They are both generated from the AK in a similar way as defined by Equation 12.1. The TEK exchange process relies on the downlink HMAC key and the uplink HMAC key to secure the exchanging messages.

Figure 12.7 exemplifies the TEK exchange process between an SS and the BS. The authenticated SS starts a separate TEK process for each SAID.

The TEK process periodically sends TEK key request messages to the BS, requesting a refresh of keying material. The BS responds to the TEK key request message with a TEK key reply message, containing the BS's active keying material. The TEK is encrypted using the appropriate KEK derived from the AK. A TEK key reply message provides the requesting SS the remaining lifetime of the keying material, enabling the receiving SS to estimate when the BS will invalidate a particular TEK and thus when to schedule the future TEK key request. The TEK process remains active as long as the SS has a valid AK and the BS continues to provide fresh keying material when the SS requires. The TEK refreshing mechanism ensures that the SS will be able to continually exchange encrypted traffic with the BS.

12.3.4 Data Encryption

Upon the completion of authentication and initial key exchange, data transmission between the BS and the SS starts by using the TEK for encryption. Figure 12.8 depicts the process, where data encryption standard with cipher

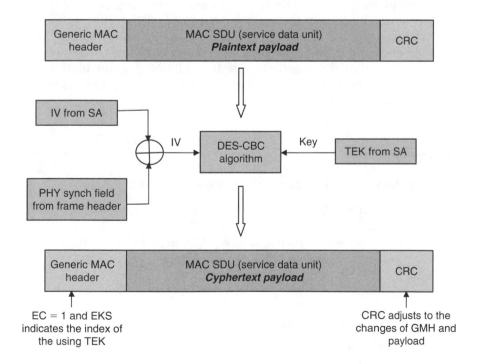

FIGURE 12.8
WiMAX MPDU encryption process. DES-CBC: Data encryption standard with cipher block changing; EC: Encryption control, IV: Initialization vector; EKS: Encryption key sequence; TEK: Traffic encryption key; GMH: Generic MAC header.

block changing (DES-CBC) encryption enciphers the MPDU payload field. Neither the header nor the CRC is encrypted to support diverse services.

As exemplified in Figure 12.8, when the security sublayer generates an MPDU, it checks the SA associated with the current connection and acquires the initialization vector (IV). The MPDU IV is generated by XORing the SA IV with the synchronization field in the PHY frame header. The DES-CBC algorithm then encrypts the MPDU plaintext payload by employing the generated MPDU IV and the authenticated TEKs. The encryption control (EC) field of the MAC header is set to 1 to indicate that the payload in the MPDU is encrypted. The 2-bit encryption key sequence (EKS) indicates which TEK is used. The CRC field is updated in accordance with the changes in both the payload and MAC header.

12.3.5 Challenges

PKMv1 uses a client/server model for traffic key management, where an SS is the client, requesting keying material, and a BS is the server, responding to the requests. The major challenge comes from the unilateral authentication. PKMv1 ensures that individual SS clients receive only keying material authorized by the BS, i.e., the BS authenticates an SS during each process but not vice versa. This implies that an SS is not capable of detecting a rogue BS. As discussed in Section 12.2, the impact of a rogue BS includes identity theft, degraded throughput, and even service termination. PKMv2 overcomes this by introducing mutual authentication, enabling the SS to authenticate the BS as well [10–12]. The mutual authentication process and other new features of PKMv2 will be elaborated in Section 12.4.

12.4   Privacy Key Management Protocol Version 2

PKMv2 is defined in IEEE 802.16e-2005 and it requires mutual authentication between SS and BS, a major deviation from PKMv1. PKMv2 also has more enhanced security features such as new key hierarchy for AK derivation and extensible authentication protocol (EAP) (see Appendix II for details) [9]. The following part of this section will introduce these significant changes.

12.4.1   Mutual Authentication

To enable mutual authentication between SS and BS, the authorization process follows these steps:

(a) The BS authenticates the client SS's identity.
(b) The SS authenticates the BS's identity.

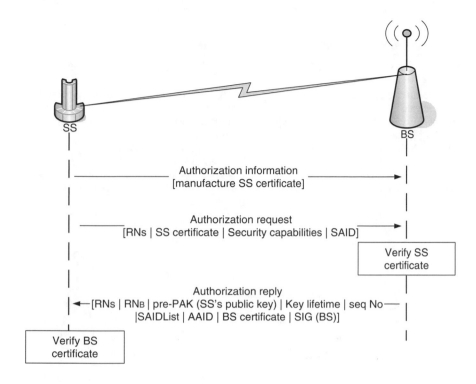

FIGURE 12.9
The mutual authorization process between an SS and the BS.

(c) The BS provides the authenticated SS with the AK, and then a KEK and message authentication keys are derived from this AK.

(d) The BS provides the authenticated SS with the identities (i.e., the SAIDs) and properties of SAs from which the SS can obtain the encryption key information for subsequent transport connections.

Figure 12.9 shows the mutual authorization process between an SS and the BS. Similar to PKMv1, the SS sends an authorization request message to the target BS, requesting an AK immediately after sending the authentication information message. The authentication information message is the same as that in PKMv1. As compared to the authorization request message in PKMv1, an SS running PKMv2 adds a 64-bit random number N_S in the authorization request message. This N_S is returned in the authorization reply message from the BS to the SS in securing the authentication process. PKMv2 also adds a 64-bit random number N_B, the BS's X.509 certificate, and BS's signature in the authorization reply message. The random numbers N_S and N_B are included in the exchange, and both the SS and BS can check the replied numbers to ensure the time freshness of the message, and thus to prevent the replay attack. Table 12.1 summarizes the contents in the authorization request and authorization reply messages.

TABLE 12.1

Authorization Request and Authorization Reply messages

Authorization Request Message	Authorization Reply Message
The manufacturer-issued X.509 certificate	The BS's X.509 certificate, used to verify the BS's identity
The security capabilities the requesting SS supports	A pre-PAK encrypted with the SS's public key
	A 4-bit PAK sequence number
The SS's basic CID, which is the first static CID the BS assigns to an SS during initial ranging	The lifetime of the pre-PAK
	The identities (i.e., the SAIDs)
	The 64-bit random number generated by the SS
A 64-bit random number N_S generated by the SS	A 64-bit random number N_B generated by the BS
	The BS's signature

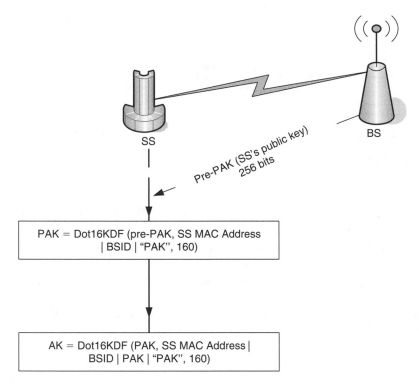

FIGURE 12.10

AK derivation in RSA-based authorization.

12.4.2 Authorization Key Derivation

The PKMv2 key hierarchy defines the key category and the algorithms used to generate keys. The authentication and authorization processes generate source key materials. These keys form the roots of the key hierarchy and will be used to derive other keys to ensure management message integrity and to

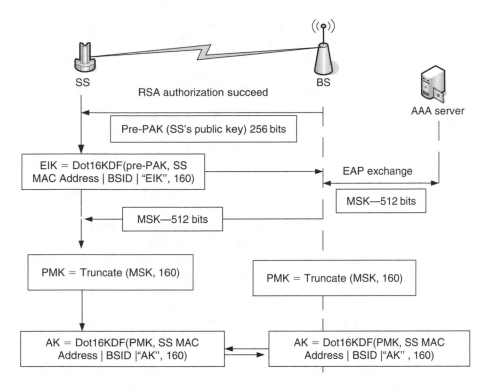

FIGURE 12.11
AK derivation in EAP authorization.

transport the traffic encryption keys. All PKMv2 key derivations are based on the Dot16KDF algorithm as outlined in Appendix IV.

PKMv2 supports two authorization schemes with mutual authentication: the RSA-based authorization process and the EAP-based authentication process. The AK will be derived by the BS and the SS from the PAK via the RSA-based authorization procedure and the PMK via the EAP-based authorization procedure.

Figure 12.10 shows the RSA-based authorization. Upon the completion of mutual authentication, a pre primary authorization key (ple-PAK) is encrypted with the public key of the SS certificate and sent to the SS from the BS. This pre-PAK is used with the SS's MAC address and the base station identifier (BSID) to generate a 160-bit PAK, which will be used to generate the AK.

In the EAP authentication mode, a 160-bit long EAP integrity key (EIK) derived from pre-PAK is used to protect the first group of EAP exchange messages. The master session key (MSK), which is 512-bit long, is the key produced from the EAP exchange. This key is known to the authentication, authorization, and accounting (AAA) server, the authenticator (BS), and the SS. Both the SS and BS derive the pairwise master key (PMK) by truncating the MSK to 160 bits at each side. This procedure is illustrated in Figure 12.11.

After EAP-based authorization is successfully performed, if the SS or BS negotiates for an authorization policy as the "authenticated EAP after EAP" mode, the SS and BS perform two rounds of EAP. After the successful first round of EAP, the SS initiates the second round EAP conversation. Once the second round of EAP succeeds, both the SS and the BS generate AK.

12.5   Advanced Security Issues in WiMAX

Although the PKMv2 protocols improve WiMAX security by adopting new features such as mutual authentication and flexible key management, there are still flaws rooted in the WiMAX standard itself.

First, since the MAC management messages are transmitted without encryption, valuable information can be given away to attackers. For example, an attacker can passively listen to the communications between an SS and a BS, intercept the management messages, verify the presence of the victim SS from the management message content, and then perpetrate a crime [5].

Second, the key management mechanism depends on the 2-bit EKS field to identify the TEK being used. The value of this field wraps from 3 to 0 on every fourth key, and thus it is easy for an attacker to interject reused TEKs [4].

Third, the original DES-CBC algorithm uses a random IV to secure the encryption, while in PKMv1 and PKMv2 the IV is generated as the XOR result of the SA's IV and the PHY synchronization field. This kind of predictable IV impairs data security. Moreover, the DES-CBC algorithm can only secure a limited length of data. It has been shown that DES-CBC loses its security after encrypting 2^{32} data blocks using the same TEK with each block containing 64 bits. Since each TEK has its lifetime, DES-CBC cannot secure data when the incoming data length during the TEK's lifetime is longer than 64×2^{32} bits [4].

As more valuable broadband services are enabled in WiMAX, more security concerns will emerge. For example, the mesh mode defined in WiMAX is more vulnerable to security threats than the traditional PMP mode. With each node being capable of forwarding traffic to its adjacent nodes, critical problems such as malicious neighbors and authorization node spoofing challenge the user privacy and system operation tremendously. Besides, secure WiMAX communications with user mobility is highly desired to facilitate seamless handoffs across different areas.

12.6   Conclusions

Driven by both the IEEE and the industrial forum, WiMAX is gaining more support from service providers as the solution for broadband wireless access. WiMAX is inevitably exposed to more security threats from the open-air channel to support both the LOS and NLOS spectra with flexible user mobility.

This chapter focuses on the PKM protocols, which play an important role to secure the connection and transmission across BWA. The processes of user authentication, key exchange, and data encryption have been reviewed with the emphasis on certificate verification, key derivation, and MDPU payload encrypment, respectively. Nevertheless, new security features in the latest standard have been covered and some open issues of WiMAX security are highlighted for future exploration.

Glossary

AAA authentication, authorization, and accounting protocol

AK authentication key, the key activated by a BS during an SS's network entry

AP access point

BS wireless base station

CID connection identifier

CPS common part sublayer

CRC cyclic redundancy check

CS convergence part sublayer

CSMA carrier sense multiple access

DES-CBC data encryption standard with cipher block changing

EAP extensible authentication protocol

EC encryption control

EIK EAP integrity key

EKS encryption key sequence

GMH generic MAC header

HMAC hashed message authentication code

IEEE 802.16 the IEEE 802.16 working group on broadband wireless access (BWA) standards, established in 1999, aims to standardize the broadband wireless metropolitan area networks (WMAN)

IV initialization vector

KEK key encryption key, the key used to encrypt the traffic encryption key

MPDU MAC protocol data unit

MSK master session key

OFDM orthogonal frequency division multiplexing, a digital modulation scheme. It splits a bit stream into several sub streams, and transmits them in parallel by modulating orthogonal sub carrier frequencies

PAK primary authorization key

PHY physical layer

PKM the privacy key management protocol defined in the WiMAX security sublayer. It protects the privacy of SS or BS via processes such as authentication and key exchange. It has two versions: PKMv1 and PKMv2

PMK pairwise master key

PMP point to multipoint

RSA Rivest–Shamir–Adleman algorithm for public-key encryption

SA security association

SAP service access point

SAID security association identifier

SDU service data unit

SFID service flow identifier

SHA secure hash algorithm

SHA-1 SHA computes a hash value of 160 bits (20 bytes) out of an arbitrary-sized binary document

SS wireless subscriber station

TEK traffic encryption key, the key used to encrypt the WiMAX traffic

WiMAX worldwide interoperability for microwave access, a technology specified by the IEEE 802.16 standards to enable the delivery of last mile wireless broadband access

WiMAX security sublayer a protocol set defined in the WiMAX MAC layer. It secures SSs, BSs, and transmission connections

X.509 an ITU-T standard for public key infrastructure (PKI)

X.509 certificate a digital certificate structured according to the X.509 guidelines

XOR exclusive-or

Appendix I: X.509 Certificate

The X.509 certificate is specified in IETF RFC 3280 (Internet X.509 public key infrastructure: certificate and certificate revocation list (CRL) profile). An X.509v3 certificate consists of three parts:

1. A certificate body containing
   - The version number (currently, v3, v2, and v1 are also possible)
   - A unique serial number assigned by the responsible certification authority (CA)

- The declaration of the signature algorithm to be used to sign the certificate
- The ID of the CA that issued and signed the certificate
- The validity period (not valid before/not valid after)
- The subject (user) ID
- The public key of the subject (user), and
- Any number of optional v2 or v3 extensions

2. The definition of the signature algorithm used by the CA to sign the certificate

3. The signature guaranteeing the authenticity of the certificate, consisting of the hashed certificate body encrypted by the CA's private key

X.509 certificates are binary encoded using the destinct encoding rules (DER). The size of a DER-encoded X.509v3 certificate containing a 1024-bit RSA public key is usually between 900 and 1500 bytes, depending on the length of the subject (user) ID, issuer IDs, and the version number being used.

Appendix II: EAP

Extensible authentication protocol is a general authentication protocol supporting multiple authentication methods such as token cards, Kerberos, one-time passwords, certificates, public key authentication, and smart cards. EAP is an authentication framework providing some common functions. It typically runs directly over data link layers such as point-to-point protocol (PPP) or IEEE 802, without requiring IP. EAP provides its own support for duplicate elimination and retransmission but is reliant on lower-layer ordering guarantees.

In the WiMAX mode using EAP, the SS sends connection request to the BS with its identity, and the BS transmits that identity to an authentication server. Both the server and the SS derive a session key. The BS gets the session key from the SS and then sends back to the server to complete the authentication.

Appendix III: RSA

RSA, named after its inventors, Ron Rivest, Adi Shamir, and Leonard Adleman, is a public-key encryption algorithm. RSA uses the private key and the public key to lock or unlock a message, respectively. The receiver sends its public key to the sender. Then the sender sends out the message encrypted with this public key. After receiving the message, the receiver decrypts the message using the private key.

The RSA algorithm consists of the following steps:

- Select two different large prime numbers p and q.
- Compute $n = p * q$.
- Select a small, odd integer $e < (p - 1) * (q - 1)$; the numbers e and $(p - 1) * (q - 1)$ must be relatively prime, i.e., they should not share common prime factors.
- Compute d, where $d * e \bmod ((p - 1) * (q - 1)) = 1$.
- The ordered pair (e, n) is the RSA public key.
- The ordered pair (d, n) is the RSA private key.

For example, given a message "M," its encrypted form is the binary message "Z."

$$Z = (M^e) \bmod n.$$

Inversely, decrypt the message "Z" by

$$M = (Z^d) \bmod n.$$

Appendix IV: Dot16KDF for PKMv2

The Dot16KDF algorithm is a counter mode encryption (CTR) construction that may be used to derive an arbitrary amount of secret key from source keying material. The algorithm to acquire HMAC is defined by the following:

```
Dot16KDF(key, astring, keylength) // generate a keylength-bit long secret key from astring
                                  // encrypted with key
{
result = null;
Kin = Truncate (key, 160);
For (i = 0; i <= int( (keylength-1)/160 ); i++) {
result = result | SHA-1( i| astring | keylength | Kin); //encrypted astring with key using SHA-1
}
return Truncate (result, keylength);
}
```

The *key* is a cryptographic key that is used by the underlying digest algorithm SHA-1; *astring* is an octet string used to generate the secret key; and *keylength* is the length of the secret key to be generated.

Truncate (key, y) is to retrieve the rightmost y bits (y LSBs) of the key only if y is less than or equal to the length of the key.

References

1. IEEE Std. 802.16-2001, *IEEE Standard for Local and Metropolitan Area Networks—Part 16: Air Interface for Fixed Broadband Wireless Access Systems*, IEEE, 2001.
2. IEEE Std. 802.16-2004, *IEEE Standard for Local and Metropolitan Area Networks—Part 16: Air Interface for Fixed Broadband Wireless Access Systems*, IEEE, 2004.
3. IEEE Std. 8802.16e-2005, *IEEE Standard for Local and Metropolitan Area Networks—Part 16: Air Interface for Fixed and Mobile Broadband Wireless Access Systems*, IEEE, 2006.
4. D. Johnston and J. Walker, Overview of IEEE 802.16 security, *IEEE Security & Privacy Magazine*, Vol. 2, No. 3, pp. 40–48, May–June 2004.
5. M. Barbeau, WiMax/802.16 threat analysis, *Proceedings of the 1st ACM International Workshop on Quality of Service & Security in Wireless and Mobile Networks*, pp. 8–15, October 2005.
6. Ernst and Young, The necessity of rogue wireless device detection, *White Paper*, 2004.
7. *CCITT Draft Recommendation X.509—The Directory Authentication Framework*, version 7, 1987.
8. RSA Cryptography Standard, *RSA Public Key Cryptography Standard #1 v. 2.0*, RSA Laboratories, October 1998.
9. B. Aboba, L. Blunk, J. Vollbrecht, J. Carlson, and H. Levkowetz, *Extensible Authentication Protocol (EAP)*, The Internet Engineering Task Force—Request for Comments: 3748, June 2004.
10. C. Eklund, R. B. Marks, K. L. Stanwood and S. Wang, IEEE Standard 802.16: A technical overview of the WirelessMAN™ air interface for broadband wireless access, *IEEE Communications Magazine*, Vol. 40, No. 6, pp. 98–107, June 2002.
11. S. Xu, M. M. Matthews, and C.-T. Huang, Security issues in privacy and key management protocols of IEEE 802.16, *Proceedings of the 44th ACM Southeast Conference* (ACMSE 2006), March 2006.
12. S. Xu and C.-T. Huang, Attacks on PKM protocols of IEEE 802.16 and its later versions, *Proceedings of the 3rd International Symposium on Wireless Communication Systems* (ISWCS 2006), September 2006.

Index